"**Outstanding . . . *The New Cool* is a book that will make you feel good about America's ability to innovate solutions to its most pervasive problems.** If you're sick of hearing about our impending doom, read this and meet a generation of kids who could change not just the nation but the world. . . . **Neal Bascomb has penned an indelible portrait of a team on a mission that may ultimately be reckoned one of the defining books of the decade.**"

> —*Dean Kamen, Founder of* FIRST, *inventor of the Segway PT,*
> *member of the National Inventors Hall of Fame, and winner*
> *of the National Medal of Technology, the Lemelson-M.I.T. Prize,*
> *and the Global Humanitarian Action Award*

"**Gripping . . . Bascomb gives us plenty of suspense and pathos. . . . An inspiring homage to the spirit of invention and a genuine sports epic, to boot.**"

> —Publishers Weekly

"**Indisputably exciting . . . the movie rights have been snapped up and with good reason . . . an unabashedly feel-good story.**"

> —Booklist

"**If I were a billionaire, I would purchase one copy of *The New Cool* for every politician in the United States, from Podunk town council member to POTUS. . . .** Simultaneously, I would commission a large Hollywood studio and a famous director to film the book with a stellar cast, just as it stands. . . . Bascomb animates his cast with a novelist's verve and deftness. They all leap off the page, fully rendered and recognizable. The dialogue is crisp and realistic, economical of purpose, with a feel for firsthand reportage. The story's suspense, while preformed due to the nature of the deadlines and the uncertainty of the contest is propelled with efficient, page-turning celerity. . . . ***The New Cool* is at once ultramodern and quintessential vintage Americana.**"

> —*Paul DiFilippo, columnist for* Barnes & Noble Review

"**I love this book.** Neal Bascomb cuts through the arcane technical jargon of robotics to reveal an amazing world of smart high-schoolers with passions for math, science, engineering, and computers. The teacher who inspires them also inspires us. **With kids like these, there is hope for the world after all.**"

—*Charles Petzold, author of* Code

"**Thoroughly enjoyable!** *The New Cool* is not just a fascinating—and dare I say, really cool—story, it also provides an important lesson. **If America is ever going to fix its broken education system and reenergize the economy, we're going to need more schools, teachers, and kids like those in this book.**"

—*P. W. Singer, author of* Wired for War: The Robotics Revolution
and Conflict in the 21st Century

"***The New Cool*** **masterfully involves the reader. . . . Bascomb captures the essence of what it means to be immersed in a real-life engineering project.**"

—*Henry Petroski, author of* The Evolution of Useful Things

THE NEW COOL

A Visionary Teacher,

His *FIRST* Robotics Team,

and the Ultimate Battle of Smarts

NEAL BASCOMB

BROADWAY PAPERBACKS

New York

BROADWAY

Copyright © 2011 by 11th Street Productions

All rights reserved.

Published in the United States by Broadway Paperbacks, an imprint of the
Crown Publishing Group, a division of Random House, Inc., New York.

www.crownpublishing.com

Broadway Paperbacks and its logo, a letter B bisected on the diagonal,
are trademarks of Random House, Inc.

Originally published in hardcover in the United States by Crown Publishers,
an imprint of the Crown Publishing Group, a division of Random House, Inc.,
New York, in 2011.

Library of Congress Cataloging-in-Publication Data is available upon request.

ISBN 978-0-307-58890-6

eISBN 978-0-307-58891-3

Printed in the United States of America

BOOK DESIGN BY NICOLA FERGUSON
COVER DESIGN BY MARIA ELIAS

10 9 8 7 6 5 4 3 2 1

First Paperback Edition

To my girls, Charlotte and Julia:

Cool is what you make it.

We are what we repeatedly do.
Excellence, then, is not an act, but
a habit.

—*ARISTOTLE*

I have never let my schooling
interfere with my education.

—*MARK TWAIN*

Societies get the best of what they
celebrate.

—*DR. WOODIE FLOWERS*

Contents

Dramatis Personae

MENTORS

Amir Abo-Shaeer A thirty-seven-year-old physics teacher and head of the Dos Pueblos Engineering Academy, he aims to revolutionize how America teaches its kids. His ragtag team of robotics rookies and their *FIRST* competition season will prove instrumental to the realization of his vision.

Stan Reifel An Internet entrepreneur and jack-of-all-trades engineer, he plays straight man to the intensely enthusiastic Amir. In the "Bat Cave" at his Santa Barbara home, the students spend long hours designing their robot.

STUDENTS

Chase Buchanan With a gaunt, angular face and surferlike calm, he is the team's lead driver. Despite his learning disabilities, he has a genius for spatial relationships that shines through during the build season, particularly in assembling the robot's drive system.

Angie Dai A quiet, retreating math savant and the daughter of two Chinese professors, she volunteered her petite frame to serve as a weight source for the team's drive-train prototype and acts as a key member of the electrical crew.

Max Garber   He is one of the team's SolidSeven, a crew that uses computer-aided drafting software to design the robot. He focuses on the robot's helix, a triumph of math and geometry, which delivers balls to the shooter.

Andrew Hsu   A classically trained cellist with the body of a football player, he becomes a master of the lathe in the machine shop. He is also one of the two lead scouts at competitions.

Gabe Rives-Corbett   Tae kwon do champion and theater-production guru, he is foremost an extraordinarily gifted computer programmer. Considered the team's "best hope," Gabe is pushed to live up to these expectations during the season and, in the process, realizes something very important about his own future.

Luke Seale   Six feet two inches tall, beanpole thin, with a penchant for wearing a beret, he designs the drive train's wheel modules. He is also the team's voice of optimism.

Stuart Sherwin   Known on the team for his thick, curly hair that hangs down past his shoulders, he is a scouting captain and member of the SolidSeven, specifically responsible for designing the turreted shooter.

Anthony Turk   The team's teddy bear, with the build to fit, Turk wants nothing more than to be the human shooter at competitions, especially since he never made the varsity basketball team. He is a fanatic Lakers fan.

Yidi Wang   A tall, bubbly concert pianist, Yidi is a member of the shooter-mechanism crew and finds herself responsible for much of its prototyping. She is also the lead presenter to the judges at competitions.

Kevin Wojcik   Earnest and one of the most reliable, hardworking members of the D'Penguineers, he is one of the team's three programmers and also the copilot of the PenguinBot.

Time Line

January 3, 2009: Kickoff Day The new game is revealed, and the forty-six-day race to build the ultimate robot begins.

February 17, 2009: Robot Ship-Date Deadline The FedEx truck arrives and the team's finished robot must be crated up and ready to be delivered to the first regional competition.

March 12–14: Los Angeles Regional Competition: Long Beach Arena The D'Penguineers compete against fifty-nine teams for the regional championship and a chance to earn a ticket to the *FIRST* Championship.

March 26–28: Sacramento Regional Competition: The Pavilion, University of California, Davis The second—and last—shot the D'Penguineers will have for regional victory and the opportunity to compete against the best of the best.

April 16–18: *FIRST* Championship: Georgia Dome, Atlanta A total of 348 teams from the United States and a host of other countries come together in the fiercest robotics competition in the world. Will the D'Penguineers be among these and earn a chance to prove themselves against the legends of *FIRST*?

Prologue

It was a beautiful morning for the championship final at Atlanta's Georgia Dome. Crisp, cloudless, a slight breeze—a perfect beginning to the day. Fans of all ages and stripes arrived at the stadium. They climbed out of the metro station. They descended from bus after bus. They traveled in packs by foot, crossing Centennial Olympic Park with a brisk step that defied the early hour. They crammed the escalators, stood impatiently in the serpentine line before the entrance, and flowed in a constant stream through the turnstiles. *Click...click...click...*the stadium attendants tallied the thousands with their handheld counters.

The fans that April morning made the usual ticket holders for a Georgia Dome showdown look bland and dispirited. Many dressed in team jerseys; the arriving crowds were a blur of colors, including every shade of neon. Some had spray-painted their hair or beards or shoes to match. Others had gone for a full Mohawk and face paint. Many walked with team flags draped over their shoulders and wore team buttons on their chests like medals.

They passed out swag by the bucket, here a rubber-chicken lan-yard, there a Weeble robot. Mascots abounded; a pink ogre strolled alongside a man in a nun's habit alongside a two-headed Martian alongside a purple-headed dragon alongside a pair of Vikings. Once in the stands, the mascots led their fans in thunderous cheers long before their teams made their way into the stadium.

All of this excitement and energy was not for the Super Bowl or the NCAA Final Four or Olympic trials or even the Little League World Series. It was for the *FIRST* (For Inspiration and Recogni-tion of Science and Technology) Championship, a high school ro-botics competition.

Down below, the locker rooms were empty. The teams needed so much room to prepare for their matches that they had overtaken an entire exhibition hall in the cavernous, glass-enclosed World Congress Center next door to the Georgia Dome. Teams were di-vided into individual ten-by-ten-foot spaces, one beside the other, in long rows. These areas were called pits, like at a race-car track, but in many the mood was the same as in a gridiron locker room before a big game. Coaches huddled with their squads, talking strategy and rehashing the scouting reports of their opponents. Some players sat on the floor, heads between their knees, focus-ing, praying, or just trying to catch their breath. Others bounced up and down, shook their arms, and rolled their necks, loosening the muscles that seemed to want to freeze them in place. "We can do this...I know it...We can do this" was a common refrain.

Some teams didn't have the leisure to worry. In their pits, they rushed to mend their robots, many of which looked like mail-boxes on wheels. Their basic shape belied their sophistication. These robots were operating with the same kind of complex, in-tegrated controls used on Mars rovers. But sophistication breeds trouble, and after these robots had been put through the wringer during the qualifying rounds, many needed repair. Before the first matches of the day, students and their coaches scrambled for the quick fix. They relinked broken chains, riveted loose frames, soldered wires, debugged code, and straightened bent aluminum rods with their bare hands. The smell of burnt metal, adrenaline, and sawdust saturated the air.

"ROBOT!" a kid shouted.

"Coming through! ROBOT!" echoed another.

One team after another called out to clear a path as they carted their 150-pound robots out of the pits. The matches were about to begin. The long walk through the concrete tunnel connecting the pits to the Georgia Dome only heightened the tension. One student, gripping the frame of his robot as though his life depended on it, said his butterflies had vanished. "I got pterodactyls in my stomach now."

Bright lights and an ear-splitting mix of cheers and dance music welcomed the players as they walked into the Georgia Dome. Teams ran through last-minute systems checks and then positioned their robots on one of the four identical rectangular fields under banners that read: ARCHIMEDES, CURIE, GALILEO, and NEWTON. These were the four divisions, and the winners from each would meet on the field in the middle of the stadium: EINSTEIN. There, at day's end, the world champions would be crowned in front of twenty-five thousand fans. The countless hours rushing to finish their robot in the six-week build period, the heated regional competitions to earn a spot in Atlanta, the brutal qualifying rounds in the Georgia Dome to have a shot at the championship—all had come down to these few final matches.

"Red team, ready! Blue team, ready!" announcers boomed on each field. "Here we go. Three. Two. One..." Robots shot forward and the matches were on. As the clock ticked down, a wave passed through the stands.

"We get what we celebrate," Dean Kamen, the founder of *FIRST,* said between matches. Given how American culture celebrates ballplayers, movie stars, and, until recently, Wall Street titans, it was no wonder to Dean that so few kids pursued studies and careers in science, technology, engineering, and math. At this world championship, and throughout the season, forty-two thousand high school students came together to celebrate something quite different.

Sporting tool belts and jumpsuits, they talked about gear ratios and computer code. They hopped up and down and screamed and laughed and cried and danced as their robots maneuvered around the field before them. They—and the thousands of fans who cheered them—were celebrating invention and intelligence.

This was a new kind of culture, the new cool.

The Kickoff

The robot is just a vehicle, just a tool.

—DEAN KAMEN

*A*t 4:30 A.M., Saturday, January 3, a white Toyota Matrix was alone on Highway 101, heading south along the Pacific coastline. Its headlights carved a tunnel through the darkness. With one hand on the wheel, Amir Abo-Shaeer drove as fast as he could without getting caught in any speed traps. He wore his usual uniform of old sneakers, cargo shorts, a black sweatshirt, and a *FIRST* baseball cap.

The son of an Iraqi theoretical physicist and an Irish Catholic from Pennsylvania, Amir was a thirty-seven-year-old teacher at Dos Pueblos High School in Goleta, California, and the founder of its engineering academy. He was tall and scarecrow thin, with neatly trimmed dark hair and an almost permanent five o'clock shadow. Small round glasses framed his deeply set brown eyes.

His wife, Emily West, a contrast of blond hair and fair skin, sat in the passenger seat. She came from a Mormon family, her youth split between Utah and California. In the backseat were two seniors from Amir's academy. John Kim had immigrated to the United States from South Korea with his family when he was twelve. His father was a professor of chemical engineering.

John spoke a stilted English and still had memories of the instructors at his Seoul grade school beating him with a bat for misbehaving. Beside him sat Kevin Wojcik, a tall, athletic seventeen-year-old of Polish descent whose father cleaned pools for a living.

This only-in-America crew was on its way to a kickoff event in Los Angeles for the 2009 *FIRST* Robotics season. The four of them would watch a live, NASA-streamed webcast of the big show in New Hampshire that would reveal the new season's game. Then they would pick up their kit of parts, the true purpose for their early-morning journey. In six weeks, Amir and his thirty-one students, most of whom were still asleep back in Goleta, a town just west of Santa Barbara, would have to build a robot ready to compete against 1,685 other teams in a game that had never before been played.

No matter what kind of game *FIRST* announced, their robot would have some essential elements, starting with an ability to move around the field of play. This did not mean legs of the C-3PO variety but rather wheeled motion. Their robot would need mechanisms to perform the tasks set out in this year's game, perhaps retractable arms or maybe a catapult. It would also include sensors, such as a meter to gauge how fast its wheels were turning or a camera to detect a target. The robot would require an electrical system that could relay information coming from the sensors and deliver energy from the battery to the motors driving the wheels and other mechanisms. This electrical system was like a body's nervous and circulatory systems combined. Finally, the robot would need a brain supplied with computer code. This brain would process the information coming from the sensors and allow the robot to both operate on its own and translate the wireless joystick commands from its drivers into action.

The kit Amir and his students would pick up would provide a starting point but no more than that. Building the robot would demand long hours and undoubtedly challenge his students as they had never been challenged before.

Unlike every other team in the *FIRST* competition, Dos Pueblos was made up of only seniors, all of whom received academic credit for their participation. They were all robotics rookies, and their season would be the capstone course at the academy where Amir was trying to create a new model for education, one grounded in real-world, project-based, interactive learning. In six weeks he needed to teach this hodgepodge group of kids how to design, machine, construct, wire, and program a robot, while bringing them together as a team for competition. Their success would prove that what he was doing in his academy was working.

Amir was always teaching, and as they drove down Highway 101, he showed Kevin and John how to predict dips in the road. He turned on his brights, pointed out a dark patch of concrete in the middle of the road ahead, and told them to wait. A second later they hit a dip. He explained that after a car hits one, any oil leaked from the engine shakes off onto the pavement. Repeated day after day by thousands of cars, this leaves a large stain on the road after a dip.

"Cool," Kevin said, but what he wanted to talk about was what kind of game Amir expected for this season. When Amir wouldn't dare a guess, Kevin and John joked that it would be some kind of contest where the robots had to swim underwater.

"No way it'll be a water game," Amir said, shaking his head and laughing. Emily sat quietly, suffering from morning sickness but saying nothing because her pregnancy, their first, was still a secret.

As they approached Los Angeles, Amir grew increasingly excited. Yes, he was eager to engage his students in this formative experience. But there was also a part of him, the one that at his age still competed in a head-to-head LEGO-building contest every

Christmas with his brother (who had a Ph.D. in experimental physics from MIT), that was personally thrilled about the new engineering challenge ahead. *It's finally here,* he thought.

As Amir pulled in to the University of Southern California campus in Los Angeles for the kickoff event, Gabe Rives-Corbett, a member of the Dos Pueblos Engineering Academy team, arrived at a salmon-colored office building near Goleta's sparse downtown. A thick coastal fog hung like a pall over the dark streets at that early hour. Because of his programming skills, Gabe was considered by some on the team to be their best hope. He had a thick frame, a round face, a mop of unkempt brown hair, a large, dark birthmark on his neck, and a grin that had *provocateur* written all over it. That grin, however, was nowhere to be seen because Gabe was still half-asleep, having awakened less than a half hour before.

Yawning repeatedly, he lugged his laptop across the parking lot and through the columned entrance of Moseley Associates, a wireless-technology company run by a teammate's father. The place was swank, with marble floors, glass doors, a grand staircase, and high cathedral ceilings. Security cameras followed Gabe as he walked down a corridor and through a set of double doors into a huge, two-story auditorium, where the rest of his team was gathered around a U-shaped table in front of a projection screen. They were a motley bunch of high school seniors: boys, girls, tall, short, big, thin, athletic and not, white, Hispanic, Asian, Middle Eastern, boisterous, sheepish, striking, plain, hiply dressed, and shapeless in hoodies and sweatpants.

Nobody was saying much, but the nervous energy in the auditorium gave Gabe a jolt. He sat down next to one of his teammates, who said, "This is awesome." As the minutes ticked down for the kickoff webcast to begin, Gabe, a former tae kwon do champion, felt the same tight pit in the bottom of his stomach that he felt when he was about to go into a fight.

Another of his teammates, Anthony Turk, who gave the impression of a teddy bear with his roly-poly form and easy, warm smile, had a different sense: fancy place, a diverse group with different skills assembled, everyone hyped to learn of their improbably complicated mission. "It's like *Ocean's Eleven*," he said. "And we're about to rob the Bellagio."

On the other side of the country, the sun was already up and a light snow was falling. Those assembled inside the Southern New Hampshire University Field House felt like kids on Christmas morning as they waited for the *FIRST* Kickoff to begin. Veterans knew there would be a lot of speeches—there was a message to spread—but the game was why most had come in person, to walk the new field, feel the new game pieces, and open the new kits as soon as possible. The NASA simulcast was great for those who couldn't be here, but for those living close enough to Manchester, or with the means to make the trip, seeing the field and everything else firsthand gave them an advantage, if slight, in understanding the new game. In this competition, teams seized every advantage they could.

High school kids and their mentors packed the field house. A stage was set on the auditorium floor with a podium, a projection screen, and two pairs of stands filled with students to serve as a backdrop for the cameras. This was a professional production: soundboards, klieg lights, a master of ceremonies advising everyone to smile and look pretty, and even someone to lead the audience on when and how to make the appropriate ruckus.

There was nothing manufactured about the crowd's eagerness, however—the shaking legs, the last-minute attempts to decipher the obtuse clues *FIRST* had provided prior to the event, and the staring at the long curtain of cloth hiding the new playfield as if by doing so they could somehow acquire X-ray vision.

The game could be anything. In 1992, the inaugural competition, four 24-pound robots had to move about a square field layered with corn kernels, collecting multicolored tennis balls of varying point values. In 1995, there was a T-shaped carpeted field with platforms and slopes where robots competed to place 24-inch and 30-inch balls through field goals. Subsequent years featured inner tubes, hexagonal fields, more balls, 50-, then 70-, then 120-, then 130-, then back to 120-pound robots, pneumatic arms, turreted shooters, human players who could score points along with the human-controlled robots, octagonal platforms that rolled on casters, ramps to mount, and tetrahedrons to stack. In 2006, robots needed to herd foam balls into corner goals as well as shoot them through a stationary 10½-foot-tall circular target. In 2008, robots ran laps around a track while maneuvering large exercise balls over and under a 6-foot-high overpass bisecting the course. At the very least, *FIRST*'s game designers had shown a Dali-esque bent for originality.

At 10 A.M. sharp, the countdown began. The lights dimmed, and the broadcast went live. The projection screen filled with snapshots of kids in various states of euphoric cheering. Then words flashed on a black screen with a *Star Trek*–style *swoosh* of sound effects:

In an earlier transformative era in American history . . .
President John F. Kennedy challenged our nation . . .
To land on the moon within 10 years . . .
Eight years and two months later . . .
Neil Armstrong set foot on the lunar surface . . .
The average age of the engineers cheering Apollo 11 that
   day . . .

Was 26...

Which means their average age at the time of the challenge...

Was 18.

Fast music blared. More snapshots of kids pumping their fists on the screen. A call to action. "Assemble your crew. Prepare for liftoff. The world is watching. Expectations are high. Duty is calling. Be the answer. One first step changed the world. Your leap can make history." The screen darkened. Big cheers followed in the stands. The lights brightened to show John Abele, *FIRST* chairman and cofounder of Boston Scientific, at the podium to make some welcoming remarks.

While he spoke about the glorious triumph of the Apollo moon landing, its fortieth anniversary that year, everyone awaited the entrance of Dean Kamen. For these students and mentors he was the star attraction, and a certain sign the game would soon be revealed.

On the morning of the kickoff, Dean Kamen awakened a little groggy in his hilltop mansion, Westwind. Early morning was not his favorite part of the day.

Usually he didn't leave his office until 9 or 10 P.M. With a few of his engineers in tow, he would grab dinner, his one real meal of the day, before coming home to do more work, whether it was mulling over a design problem, tinkering in his shop, or talking with one of his frequent guests, many of whom ended up staying overnight. Dean wouldn't go to bed until he was so tired that he knew the second he slipped under the covers he would be out. No sense in wasting minutes counting sheep. The sole difference the night before the kickoff was that he had left the office early to host an event at his house for *FIRST*. The night went typically late.

Dean treated his few hundred guests, mostly team mentors

from around the country, to a boy's dream of a home. These teachers and engineers, the bedrock of the *FIRST* community, looked about in wonder, uttering "Wow" and "Cool" and "Sweet." At 20,000-plus square feet, it was the biggest house most of them had ever seen, let alone been inside. More amazing, Westwind was one of a kind, a shrine to the magic of invention.

The hexagonal house was set on a steep hill, the highest spot Dean could find between Manchester and Boston. Its centerpiece, the form around which Dean designed the house, was a steamboat engine previously owned by one of his heroes, Henry Ford. The 25-ton, three-story marvel, a beautifully curved hulk of metal that Dean had rebuilt, bolt by bolt, was proof that engineering was art. The house, a labyrinth of rooms, levels, and passageways (some cut straight out of the hill's black rock), muddled one's sense of direction.

Getting lost long enough, a visitor might discover a professional-grade machine shop; an indoor swimming pool; the elevator from the set of *The Sting;* countless bedrooms and bathrooms; a chess-playing robot named the Turk; the first flight simulator; a 1913 Model T; jukeboxes; slot machines; Einstein portraits; a hangar with two helicopters; an exercise room; a sauna; a garage housing a black military Humvee, a Porsche 928, an electric Tesla Roadster; a cupola from which to watch the weather shift or planes taking off at the local airport; two pulley systems to deliver bottles from the wine cellar; and a secret corridor leading from the wood-paneled library, opened by pulling at the spine of *Ingenious Mechanisms for Designs and Inventions,* volume 4. A 150-foot windmill powered the house, and a 3.8-million-gallon pond, which Dean had created by blasting out rock and dirt from the hillside, supplied his water. For physical activity, the estate boasted a floodlit baseball diamond and tennis court.

Some likened Dean to Willy Wonka. He possessed the fantastic house and more than enough of the eccentricities. The morning of the kickoff, he put on one of his more than two dozen identical outfits: blue jeans, a blue-denim shirt, and work boots. There was a tuxedo in the closet as well, but he hadn't even worn that when he attended an event at the White House. From his bedroom, he walked down a spiral staircase and through a doorway into the kitchen to have his coffee. An overnight guest waiting for him there would have been puzzled to see Dean emerge from the cupboard, as if he had spent the night cramped inside. Dean had the door leading to his bedroom designed to look like his kitchen cabinets.

With a light snow falling, he planned on using his Hummer to get around that day. Usually he flew, taking his "magic carpet" to work, an Enstrom helicopter he had modified. The *Star Wars* theme music played in his headset every time he lifted off. The journey to the rooftop of his company headquarters in downtown Manchester took three and a half minutes, the *whoosh, whoosh, whoosh* of the helicopter blades signaling his approach. On occasional weekends he'd fly out to his private, 3-acre island off the Connecticut coast, which he had named North Dumpling Island. He had whimsically declared the island's secession from the United States, printed his own currency and newspaper, written a constitution and national anthem, deemed himself Lord Dumpling, and proclaimed that he and his citizen-guests had created "the Only 100 Percent Science-Literate Society."

At fifty-seven, Dean lived alone, had never married or had children. His chocolate factory was the DEKA Research and Development Corporation, his Oompa-Loompas the hundreds of engineers working for him. But this was where the analogy broke down. Willy Wonka, Dean was not. There was nothing frivolous or whimsical about his inventions or clownish about his engineers, many the best and brightest in their fields, who labored long hours for less pay than they deserved, all to work hand in hand with a visionary.

There had never been any doubt about the career Dean would pursue. At the age of three he had awakened his father, an artist and prolific comic-book illustrator, asking him to sketch out an invention Dean had thought up in the middle of the night. "Put a toggle switch here, make the box shorter, sketch an outlet for a wire," Dean had instructed, his father feeling like a police artist trying to get on paper what his son had in his head. A few years later in his Long Island, New York, home, Dean had rigged the blanket on his bed with ropes and pulleys to make his bed instantly. Even then he seemed in a hurry.

A teacher at school called him "Deano the Relentless." He was always questioning the nature of things. For instance, what causes a mug of hot water on a table to cool rather than heat up further. When his teacher couldn't provide sufficient answers for his never-ending questions, Dean started seeking them himself. At thirteen, he tracked down a copy of Sir Isaac Newton's *Principia* to understand the laws of physics, and it was like the world opened up to him for the first time. He later said, "$F = ma$, that's a pretty simple linear equation. Force equals mass times acceleration. You can't come up with a more simple statement that's not trivial. Yet it describes the motion of billiard balls and of galaxies. On a galactic scale, you can predict where the *next eclipse* will be because of $F = ma$. You can predict where *atomic collisions* happen because of $F = ma$. That's *astounding*."

From that day forward, he stopped relying on his teachers to explain the laws of nature to him. On his own, he read Newton and Einstein and others. School left him unsatisfied, and tests seemed ridiculous. He made only passing grades. At sixteen, he started creating electrical contraptions that could make the lights in a room pulse to the beats of music. These inventions led to an assignment to build a light show at the Hayden Planetarium in New York. Soon after graduating high school, he was making sixty thousand dollars a year building automated control systems for light-and-sound shows.

He went off to Worcester Polytechnic Institute, but even there the classes left him unchallenged. One day his brother, who was in medical school, told Dean that he should invent a small pump to deliver drugs to patients intravenously. Dean assembled a machine shop in his parent's basement and created a prototype that received favorable mention in *The New England Journal of Medicine*. While his parents were away on vacation, he had their Long Island house jacked up on stilts so that he could extend their basement to give him more room for his mini factory. Then he started creating a line of miniature wearable pumps, including one to deliver insulin to diabetics. His younger brother became his assistant, his mother the accountant. Eleven years later, at age thirty, he sold AutoSyringe to a medical-device manufacturer for millions. He moved to New Hampshire and started DEKA (a play on his first and last name). As for Worcester Polytechnic, he never bothered graduating.

At DEKA, he invented kidney dialysis machines—one no bigger than a briefcase—to replace the refrigerator-sized contraptions in use at the time. Like his portable drug-infusion pumps, it obviated the need for hospital trips. Then there was an intravascular stent that improved blood flow to the heart, the motorized wheelchair that allowed the disabled to climb stairs (the forerunner to his infamous Segway Human Transporter), and recently a lightweight prosthetic arm that was sensitive to pressure.

With each year, the problems Dean strove to solve were bigger in scope and ever more intractable. They included halting global warming; providing everyone in the world, regardless of location or circumstance, with potable water and electricity; and ridding cities of traffic congestion and pollution. With DEKA, Dean was making some progress on these problems, but no matter how smart he and his engineers were, how ingenious their ideas, how industriously they worked on them, Dean knew it was not enough. That was why he considered *FIRST* his most important endeavor, and why he was frustrated it was not growing fast enough.

The idea had come to him one rainy Saturday in 1989. Because of the weather, Dean had driven to DEKA headquarters, located in a former textile mill on the banks of the Merrimack River. He found the parking lot crowded from the visitors to the hands-on science center that he had built soon after moving shop to Manchester. Instead of going up to his office straightaway, he went inside the center.

It had always perplexed Dean that the scientific world that so fascinated him was lost on most people. He was not alone in that feeling. In 1959, when Dean was eight years old, a British chemist and government official named Charles Percy Snow authored *The Two Cultures and the Scientific Revolution.* He wrote, "A good many times I have been present at gatherings of people who, by the standards of the traditional culture, are thought highly educated and who have with considerable gusto been expressing their incredulity at the illiteracy of scientists. Once or twice I have been provoked and have asked the company how many of them could describe the Second Law of Thermodynamics. The response was cold: it was also negative. Yet I was asking something which is about the scientific equivalent of: *Have you read a work of Shakespeare?*"

This divide in modern culture had been one of the motivating forces behind Dean's decision to launch his science center. Seeing how crowded it was that morning, he felt pretty good about himself. He had built something real here, something that was showing kids how fascinating science and technology were.

Then a young boy wearing a Celtics cap ran past him. Dean

started looking around, noticing how many of these kids were wearing sport caps or jerseys, the names of their favorite Red Sox, Bruins, or Celtics star stitched across their backs. He went up to one kid and asked him to name a famous living scientist or inventor. Nothing. Dean asked the next kid, then the next, and then their parents. Finally one adult offered, "Albert Einstein, but I think he's dead."

Dean felt his heart sink into his shoes. Pride in his center gave way to despair. He wasn't changing any lives. Everyone had come here because the admittance was free and it was raining. Tomorrow when the clouds cleared, these same families wouldn't think of coming to the center. They'd rather pay fifty bucks to see a baseball game, and then wrap up their day by watching their favorite action star in his latest film. Dean felt he had been naïve to believe this center would change that.

He went up to his fourth-floor corner office and stared out the window to the river below, the thoughts coming in a rush. Kids revered athletes and movie stars, wanted to grow up to be just like them. They were the only heroes these kids knew, and American culture kept selling to them that this was okay. "Be Like Mike," the billboards called.

Dean decided he needed to create something that made kids want to excel in thinking, something that truly showed how thrilling science and technology could be. He needed to invent a sport that was about smarts, that had its own set of heroes, its own hall of fame, its own supporters and coaches, its own Olympics. Frustrated, incensed, and excited all at once, Dean formed *FIRST* the same day.

A few months later he met Dr. Woodie Flowers, and the sport took shape. Flowers, a pioneering MIT professor of mechanical engineering, invited Dean to lecture to his graduate design class. As Flowers remembered, the two had a "philosophical love-in" about the need to change culture. He soon visited Dean in Manchester to see what they had cooking in the DEKA labs. At one point, he showed Dean a documentary about his class Engineering Synthesis and Design, known ubiquitously on campus as 2.70. The class centered on a design challenge where students had to build a machine out of a kit of parts to accomplish a specific task. Woodie had turned this class competition into a university event that was more popular than its football games. "Great," Dean said when the documentary was over. "Let's do it." In 1992 they held their first competition.

In its eighteen years, *FIRST* made huge strides. From twenty-eight teams of high school students who participated in the inaugural competition in a New Hampshire high school gym, it now boasted more than fifty times that number of teams. *FIRST* also offered the Junior *FIRST* LEGO League for six- to nine-year-olds, the *FIRST* LEGO League for nine- to fourteen-year-olds, and the *FIRST* Tech Challenge for high school students. In total, the programs racked up some incredible numbers: 212,000 students, 19,000 teams, 57,000 mentors, and 33,000 event volunteers. The organization had more than 3,000 corporate sponsors and participants were eligible for more than $10 million in scholarships in 2009.

Its impact was clear. A study by Brandeis University compared *FIRST* and non-*FIRST* high school students from similar backgrounds with similar experiences in math and science. Those who participated in *FIRST* were more than twice as likely to pursue a career in science and technology and four times as likely to pursue one in engineering.

Why was *FIRST* working? Woodie Flowers understood it better than most. "I think it's time to shift away from the 'Sage on the Stage,'" he explained. "What do we do? Learn by doing. First, understand the phenomenon before we tackle the abstractions...."

Second, learning *with* always trumps learning *from*." For too long, Woodie believed, schools have focused on what he called "training" instead of education. In a speech at Olin University, he broke down the difference:

> Training and education are very different. Training is a commodity. Education is the part that confers comparative advantage. Much of what we call engineering education is in fact training and poorly done. Learning calculus is training. Learning to think using calculus is education; learning spelling and grammar is training. Learning to communicate is education; learning a CAD (computer-aided design) program is training. Learning to design is a much more complex, sophisticated thing; learning the parts is training. Learning the synthesis and whole is education. It's not clean. The boundary is clearly fuzzy. Once you could be trained to be a professional if you knew things, that was enough, but information is ubiquitous, you can't have an advantage in society because you know something.

In Woodie's breakdown, the robotics competition fell on the side of education. But despite the clear impact and rapid growth of *FIRST*, Dean was desperate to do more. *FIRST* was in 8 percent of high schools across America. He wanted to be in every single one.

Dean feared the United States would become a "banana republic" if it didn't change things around quickly. He knew the statistics. By their senior year in high school, a mere 18 percent of students were considered proficient in math and science. Among forty-nine industrialized countries, eighth-grade American students ranked in the middle of the pack on science and math tests. By twelfth grade, they ranked at the bottom, above only those of Cyprus and South Africa.

American universities graduated 70,000 engineers in 2004, while China graduated 500,000 and India 200,000. Only 32 percent of students in the United States graduated college with science or engineering degrees, while China boasted double that percentage. Overall, American students were failing to pursue the fields of science, technology, engineering, and math (STEM) in great enough numbers, and the statistics were even more abysmal for women and minorities.

If the United States did not do everything in its power to promote STEM careers, then the country would fall far behind its competitors around the globe, threatening both its prosperity and security. The country had known this when the government invested in science and technology after the Russian launch of the *Sputnik* satellite in 1957, but Dean felt America had lost its way since. Wall Street had not made the country an economic superpower, he reminded everybody. Its inventors, scientists, engineers, and entrepreneurs had. "America's only going to have a high standard of living if we remain productive," he said. "We're in danger of giving away the goose that lays the golden eggs."

Dean wanted America to flourish, but he was focused on the bigger picture as well, evidenced by his activities at DEKA. He wanted more engineers and scientists so that a greater army would be battling the challenges that the whole world faced. Only when these were solved would he feel like he could rest easy and enjoy those vacations he never took.

The morning of the 2009 kickoff he steered his black Hummer down the icy road from Westwind, energized, despite his lack of sleep, at the thought of seeing the kids who had answered the call of a different kind of sport. They remained his greatest hope.

A half hour into the kickoff, anticipation turning to impatience in the stands, Dean Kamen arrived on the stage to a roaring ovation. He was five feet six, but his pompadour of thick black hair gave him a few more inches. He had the wiry body and lean face of someone who had to be reminded to eat. Looking out into the stands, he began his speech, no notes, no teleprompter.

He was not a great orator. With his faint Long Island accent, he tended to swallow his words instead of projecting them. He wore a half smile on his face, gestured a lot with his hands, and crossed his feet underneath the podium, the sum of which made him seem uneasy. Instead of building to a crescendo, his speeches rambled and then tapered off at the end, almost like some internal clock had told him he had gone on too long. Still, there was tremendous power in his words because it was plain that he believed in what he was preaching, lived it to his core, and was now asking for the best that these teenagers had in them, a challenge most had never been given before.

"Why do we do *FIRST*? Because the world's a mess. Read the news. Look around you. We got lights, clean water, ways to get around. We have hospitals, schools, safe malls. But two-thirds of all people alive today, 4 billion people, live on less than $2 a day. Half of them live on $1 a day. That's their whole life. We're the richest in the world, by far. And the world's a mess. Somebody's got to fix it. Do you think the people living on a buck a day, who don't have clean water, schools, technology, education, do you think those people can fix it? No. *You* have to fix it.

"I started *FIRST* because so many kids in this country were ignoring that opportunity. Our culture has convinced so many that knowing about Britney Spears or Paris Hilton is more important than real knowledge. We've got to find a way to convince all kids that we have to start celebrating the stuff that matters. We not only have the opportunity, I think we have the obligation—the moral obligation—to help the rest of the world.

"So why do we do *FIRST*? I think it's easy, because the world needs a lot more technology quickly. You've got to be better at it than we were, you've got to learn faster, and you've got to deliver at an earlier age. If you want exciting careers, you better start remembering what used to be true: We have to create wealth. For a while there, the bankers were doing a pretty good job of moving it around. We've got to start creating it again. People who create solutions for problems with the tools of physics and engineering are the people that we will count on. To rebuild our economy. To rebuild the world to make it better than it was. Why do we do it? I think it's pretty easy.

"The robot is just a vehicle, just a tool. The skill sets you walk away with will give you careers for a lifetime. *FIRST* is a genuine card-carrying microcosm of the real world of engineering. We give you a little time, never as much as you need. We give you a little bit of material, never what you'd really want. You never know what the competitors are doing. *FIRST* really is a way to show you what the world of science, technology, inventing, and problem solving is. It's all hard, and if this frustrates you, tough, it's important."

Heady and direct stuff for a young crowd, but for a moment, they forgot about the raising of the curtain and were inspired. But as Dean went on, revealing at one point that the new game featured a "slippery slope," eyes drifted back to the curtain, feet shuffled, and there were a few sighs, all of which seemed to say, *Let's get going.*

Then Dr. Woodie Flowers came out, and people sat forward in their seats again. With a bushy mustache, long gray hair tied back in a ponytail, and a big *it's all good* smile, he looked like he had teleported in from the 1960s.

"Good morning, all," Woodie said with a liveliness that Dean rarely showed in front of large crowds. Woodie gave his own short speech about how everyone there could be his "own individual economic stimulus package," then quickly moved on to introducing the game.

A *FIRST* staff engineer crossed the stage to where Woodie was standing. He towered over her. They clasped hands and started trying to yank each other off balance, a struggle he soon dominated. "On Earth, if I were having a tug-of-war with Kate, I'm heavier than she is, I'd probably win," Woodie said.

A heavy rope then descended from the ceiling, one end an empty loop, the other tied around some gym weights. Assistants pushed a large wooden crate onstage. After climbing onto the crate, Woodie looped the rope around his chest and stepped off. Instead of crashing to the floor, he glided down.

The MIT professor was in his element, enjoying the demonstration, while the students and mentors clasped their hands in their laps, trying to figure out what all of this meant.

"But if I were on the moon, and five-sixths of my weight were being offset," Woodie continued, "I could jump higher, like those guys on the moon." And he jumped into the air like he was in a zero-gravity chamber. "But in this tug-of-war, I don't have a chance." Kate grabbed Woodie and yanked him toward her, easily. "She has much more traction than me. So what do you think this silly demo has to do with this year's game?"

A drumroll started. Dean and Woodie disappeared behind the curtain. The audience was leaning so far forward in their seats they looked like they were about to fall over.

"And now," the emcee said, "the moment you've all been waiting for. Ready. Go." The curtain drew aside to reveal a large rectangular field. The hushed silence in the stands was replaced with a collective "What?"

A shiny white material covered the field, instead of the carpet of previous playfields. More perplexing, the field was empty except for six strange hexagonal bins filled with even stranger balls. Typically the field had a track or ramps and a complex arrangement of game pieces that the robots would have to maneuver.

"Wait a minute. There's nothing here," Woodie said, his words obviously from a rehearsed script. "How are we going to explain the game? Roll it."

All eyes focused on the projection screen. The auditorium darkened, and a narrated video began, using Pixar-like animation to demonstrate the game. "Welcome to the 2009 *FIRST* Robotics Competition, and this year's game, Lunacy."

Three minutes later, the lights came back up to reveal a room full of blank faces. Dean and Woodie further stunned the crowd by bringing out two boxes that they said each weighed 150 pounds. Each box was set on four wheels that appeared to be made of the same material as the field. Dean and Woodie pushed these boxes around with little more than nudges of their feet. It was clear that this special surface was as slippery as promised in the narration.

Then the students and mentors were let loose on the field. They could not have rushed any more if someone had screamed, "Fire!" On their way, they reached for their digital cameras, cell phones, video cams, rulers, notepads, and pens to record their observations. Kids first went for the woven balls. They were stretched, compressed, rubbed against sweaters to test for their adhesive qualities, and thrown into the hexagonal basketlike bins.

They understood that the bins, called "trailers" in the video, had to be hitched to the back of each robot. A few kids circled these trailers until one mentor went in for the kill. He upended a trailer's low platform to check out its two slick white wheels, which their robot would also have to use. He examined the wheels with the attention of a jeweler eyeing a diamond. Others made the same close investigation of the playing surface. On hands and knees, students ran their fingers across the strange white material, called regolith, which felt like Formica (hard, slick, and synthetic). They measured everything on the field: ball diameters, trailer dimensions, and the railings and clear walls surrounding the playing surface.

Within five minutes a mass of people covered the field. Whispered conversation rose to a roar as they floated ideas about what to do, what strategy to take. "Oh, this is key," remarked a student while she used a ruler to measure the narrow band of carpet that ran around the edge of the field. "Look at this," another said, throwing a ball so that it bounced off the round post in the middle of the trailer like a backboard.

Woodie Flowers made his way toward the field, a gaggle of students and mentors surrounding him looking for any tips. He just smiled, coy about what kind of robot would best serve the game. Dean too kept mum as he stood among a host of eager competitors, though he remained some distance from the field, where one risked being crushed by the crowd or struck by one of the balls flying through the air.

There was still a game manual the size of a telephone book to read, and the kit of parts to open, but already a massive amount of brainpower from forty-two thousand students and more than eight thousand mentors, including the Dos Pueblos Engineering Academy team and their teacher, Amir, was focused on understanding this new challenge. On February 17, forty-six days from the kickoff, they needed to have mastered the challenge and shipped off their finished robots in wooden crates to their first competition. The deadline loomed.

Strategy

We're not going to be pulling any punches this year. It's either go big or go home.

—AMIR ABO-SHAEER

In a lecture hall at USC, Amir stared at the projection screen after the narrated video ended. A big *wow* spread across his face.

"This is going to be crazy," he finally said to Kevin and John. The game looked like basketball meets robot tag to him, and he wanted to grab the kit and return to Goleta as soon as possible. They scooted out of the lecture hall, Amir setting the pace with his long, quick strides. Before other teams had made it to the line to get their kits, Amir was already speeding out of the campus, the back of his Matrix weighed down with robot parts. The season for the Dos Pueblos D'Penguineers, Team 1717, had started.

Kevin, who was as earnest as they came, opened his laptop and punched in the password provided during the kickoff to open the Lunacy rule book.

"This is going to be hard," Amir said to Emily. "Harder than other years."

In the backseat, John and Kevin shared a scared look.

As Amir fired off questions about the rules—robot dimensions, wheel requirements, the different types of balls in play—he was overwhelmed by what was at stake this season.

He didn't want to let the students down. As seniors, this was their one shot at the competition. For many this would be the closest they'd come to a sporting experience in high school. He had to help them achieve their best. He also didn't want to let their parents down. They had entrusted their children's education with Amir when the engineering academy was still an unknown. They had donated their time and money to help create it.

But Amir felt this obligation to his students and parents every year. This season was different: The future of the academy depended on their success.

Almost a decade before, Amir had left a career in mechanical engineering to teach. He already had a master's degree in engineering, and he could have finished his Ph.D. to teach at the university level. Instead he earned his master's in education to become a high school science teacher. He felt like he could impact students more at that time in their lives than any other.

Soon after he started teaching physics at his alma mater, Dos Pueblos, he realized the challenge he faced in keeping his students engaged. The prescribed textbooks and required battery of tests were obstacles more than anything else. Every class seemed as much motivational speech as lesson, and he left each one exhausted.

His students approached his class, as he was sure they did others, like it was merely another stepping-stone they had to cross on their way to graduation and then on to college. They knew they could find online most of what they were being taught in their classes. A thread on Wikipedia or a video lecture on YouTube offered the same material—often more compellingly. They also didn't see the relevance of what they were having to learn or, worse, memorize.

But Amir knew he and his fellow teachers had an opportunity

with these kids. For six hours a day, five days a week, thirty-six weeks a year, these kids were a captive audience. Instead of shoveling them information available elsewhere, teachers could give their students experiences that would engage them on multiple levels and prepare them for whatever future they pursued, experiences that couldn't be found in books or on the Internet.

Fate handed Amir a chance to try this alternative approach. When he started at Dos Pueblos, the school had just received a grant to start a precollege engineering program. They had an outline for what kind of program they wanted to build, but they didn't have anybody with an engineering background to run it. The assistant principal came to Amir in his first semester and offered him the post. Amir buried himself in the project. Working weekends and throughout the summer, he created a curriculum that he hoped would excite students and invest them in their own education.

In 2002 he launched the Dos Pueblos Engineering Academy. It would be a four-year program, with thirty-two students per grade level. Along with their core requirements, students would take one or two courses a semester in the academy, starting with introductions to physics, engineering, and computer science, followed by more advanced courses in each. All these courses balanced lectures with individual and team projects.

These projects—whether building underwater remote-control vehicles or Rube Goldberg contraptions that moved marbles down ramps and up elevators—connected academics to the real world. The underwater vehicles brought to life lessons on buoyancy and drag, electric power and motors. The Rube Goldberg contraptions illustrated potential energy, friction, and mechanical advantage.

Further, the projects taught them analytical thinking and problem-solving skills. They engaged the students to compete against one another as well as to work together toward a common goal. Regardless of their future careers, these projects provided lessons that would serve them throughout their lives.

When the academy started, Amir knew he needed a capstone course for his seniors, something that brought home and expanded on everything they had already learned. For a long time, he waffled between creating his own internal engineering project and connecting with an external project already in place.

Then one day an engineer at Raytheon who was helping Amir develop his academy curriculum sent him a note about a robotics competition called *FIRST*. The name meant nothing to Amir, but the idea of robotics excited him.

Several years before, Amir and Emily had honeymooned in British Columbia after they graduated together with their master's in education. At the hotel, he had watched an A&E special that featured a high school robotics competition. He had been inspired by its educational potential and thought that this was just the type of experience that would excite his students. He hadn't known it then, two weeks before his first day as a teacher at Dos Pueblos, but he had been watching a *FIRST* competition.

After the tip from the Raytheon engineer, Amir learned everything he could about the organization founded by Dean Kamen. He attended a *FIRST* seminar and took some students to see a regional competition. Blown away by the possibilities, he formed a team from his first class of seniors at the academy for the 2006 season. They were the 1,717th team to register for *FIRST,* thus their numerical name.

Although Amir had no experience building robots, no machine shop, no spare funds, no place to build, and nobody to help him, he forged ahead. He procured a mill and lathe that he placed in a closet across campus from his regular classroom. Over Christmas break before the first season began, he rewired the closet. The team used his classroom to do most of the building and the makeshift machine shop to create the parts they needed. Amir skateboarded back and forth between the two to keep everyone on track. He worked twenty-hour days. He slept on the classroom floor on a cardboard box like a hobo. Although the team lost most of their matches at their first regional competition, they won the Rookie All-Star Award. This earned them a spot at the *FIRST* Championship in Atlanta.

Their first season was a big success, but nobody paid much attention at Dos Pueblos or in Goleta. Before the next season, Amir tried to get some space on campus other than his classroom to build their robot. School officials denied his petition. His new team performed even better at competitions than the one the year before, though this time they didn't make it to Atlanta. Again, Amir tried to find space for the team. Again the school denied him, even when he learned that a new $15 million theater would leave the old one all but abandoned.

At this point, the engineering academy had earned a sterling reputation: Its students' test scores were off the charts; kids transferred to Dos Pueblos to join the academy; and Amir was gathering accolades and Teacher of the Year awards. Every year, three times more freshmen applied to the academy than Amir could accommodate. These numbers drove Amir insane because he knew he was on the right path and students were being left behind. He needed additional teachers for his academy, more classrooms, and better facilities. Yet he couldn't even get a simple space big enough for his robotics team at a school with a new theater, an impressive football stadium, and a recently completed $5.9 million Olympic-sized outdoor swimming pool. The situation demoralized him.

In the summer of 2007, Amir learned that the State of California offered grants to school districts to enhance career technical education. Districts could apply for up to $3 million, which was great, but whatever funds they were granted, they had to raise a matching amount on their own. Amir had never tried to raise this kind of money, but that did not stop him from deciding that he should aim for the maximum grant amount.

Next challenge: The deadline for the grant proposal was a month away. Most districts had whole committees working for a year on their proposals. Some had even hired architectural firms to draw up plans for what they would do with the grant. Amir didn't care. The grant was his only chance. To expand the academy, he needed a building and funds.

He and Emily worked night and day on this project. Together, they wrote and assembled a fifty-page proposal, gathered endorsements, and put together a budget. Amir even sketched the plans for his own building. By the end, they had spent so many hours on the proposal that Amir thought for certain he would get the grant. In November, he did.

Raising the matching $3 million proved the next challenge. He assembled the parents of current and former students and told them about the grant. They decided to start a foundation to raise the money, and several of the parents signed on to help run it. They created donor lists and presentations about the academy. Amir was as much entrepreneur as teacher now. He understood that if he wanted to change the face of education, he would have to push the fight forward himself.

Architects drew up shovel-ready plans for a 12,000-square-foot facility. In meetings with the foundation board and donors, Amir began to realize the opportunities they would have with such space—and with substantial funding. Students could do research-and-development work for local companies. He could use the building for summer camps and outreach programs. Best of all, the academy could become a teaching center. His program met the state-approved curriculum for high school students, no small feat, and teachers would be able to come and learn how to develop similar academies throughout California and beyond. Everything depended on the grant, however.

The year 2008 went well. After the grant announcement, the school found a small space for the robotics team. They won two regional championships, and the foundation raised half a million dollars.

Still, by the start of 2009, Amir needed an additional $2.5 million to secure the grant. If he failed to raise these matching funds by the November deadline, the grant would disappear, and with it his dreams for helping change the course of education in America.

But he was not deterred, even with the collapsing economy and drying-up donations. None of what he had done had come easy. If he had quit every time he was told "No, you can't do that," at every financial or bureaucratic roadblock, his academy wouldn't have been more than a few elective classes at Dos Pueblos.

With the approach of the grant deadline, the pressure was mounting. Every week Amir bought a lottery ticket with the idea of giving the money to the academy. He knew the long-shot odds, but sometimes it felt nice to put one's faith in the hands of fortune.

His real hope rested in his robotics team, the D'Penguineers. They were the best and most visible example of what he was trying to do with his academy. Some considered their successful 2008 season to be a fluke. Amir needed to show repeatability, or, he feared, all the momentum would be lost.

The problem was, of all the classes that had gone through the academy, this one had proven the most divisive. The group had a "bad spirit" about them. They knew it. He knew it. And this was the team that would determine the academy's future.

As he drove out of Los Angeles, the overwhelming feeling of what was at stake was replaced by the familiar excitement of tackling a new engineering challenge. He kept his eye on the rearview mirror for tailgaters, a constant concern. He had always wanted a bumper sticker that said, IF YOU CAN READ THIS, YOU DON'T UNDERSTAND PHYSICS.

Amir called Gabe and told him to inform the team that they were not to speak about what kind of mechanisms they wanted on the robot because it might channel their thinking too early. "Don't pollute the well. Only talk about the rules and strategy to win the game, how it's supposed to be played. Okay? I'm on my way."

An hour before at the Moseley Associates auditorium in Goleta, the Dos Pueblos team had been wondering when the drone of speeches would end. Gabe had fallen asleep and was nudged awake when the curtain was about to be drawn on the playing field. "This is it," someone said in the dark. "They're going to show it."

Gabe stared at the screen as the soft, soothing voice of the animation's narrator filled the auditorium. Everyone around him was stone still as well. They all knew that what they were about to hear would rule the next six weeks of their lives:

"On a 27-by-54-foot field known as the crater, alliances of three teams are each positioned in bases at each end of the field where they control their robots."

The video opened onto a white field. Around its perimeter stood six stationary robots (three red, three blue) hitched to basketlike trailers. Outside the field, standing behind tall clear walls, were a bunch of blue and red Gumby-like figures, symbolizing people. The video panned around the field, and Gabe went dizzy at the assault of information.

"The crater is covered in a slick polymer material called regolith. It provides a unique surface for the robots to drive upon. Special

wheels are used on the robots to create a low-friction interaction with the regolith. This simulates the low-traction effects of driving in one-sixth gravity on the surface of the moon."

A close-up showed a wheel burning out on the regolith surface, and Gabe knew this was a real danger. Like a tire on ice, the wheels would slip if they tried to accelerate too fast. Conversely, it would be tough to stop.

"Lunacy uses three types of game pieces, known as moon rocks, empty cells, and super cells. A trailer is attached to each robot. The trailers are the targets for the opposing alliance. The objective of the game is to get as many moon rocks and super cells into opposing trailers as possible."

Three woven balls of different colors spun on the screen. A trailer slid across the field, and balls plopped into its basket until they spilled over the side. Gabe, who had done some research into previous seasons, knew that this was typical of *FIRST*: different game pieces worth varying point tallies to multiply the strategies to win. Alliances, where their team would be paired with two others, for three-on-three matches, were also typical.

"At the start of the game, human players, known as payload specialists, are in position behind opposing robots. Each robot is placed with a trailer touching the wall of the crater. Robots are able to start the game with as many as seven moon rocks in their possession. During the autonomous period, players attempt to launch moon rocks over the walls into opposing trailers."

This is bizarre, Gabe thought—and not just because of the space-mission theme. A figure situated behind each of the six robots tossed balls into their trailers as soon as the buzzer sounded for the first stage of the match, autonomous. Red players shot into blue trailers, blue players into red. Even as the robots moved away, following preprogrammed instructions to avoid attack, the figures continued to arc shots over the walls and into the trailers. They were making a lot of shots, leading Gabe to believe that human players would take a more significant role in this game than in other years. There also seemed to be numerous balls in play.

Across the table, Chase Buchanan, who had long, straight, brown hair, a gaunt, angular face, and surferlike calm, started chanting, "Turk…Turk…Turk."

Everyone turned to Turk, who had a silly grin on his face. As usual he was wearing a Lakers shirt. He had never made the cut for the high school basketball team, which they all knew he wanted more than anything. Now here was a game where they needed "payload specialists" to make jump shots. "It's just like basketball. I can't believe it," Turk said, breaking the tension. Nearby, Bryan Heller, who was on the varsity basketball team, was quiet.

"Or robots may use their cameras to track each robot trailer; this allows them to score in their opponents' trailers."

The robots, all with square bases and absurd contraptions on top—for example, a tennis racket to propel balls—maneuvered about the field. The robots showed how the game would be played but were so cartoonish that they offered little hint of what kind of robot the teams should build. One had a short cannon with a human eye above it. This blue robot hunted down a red robot and launched balls into its trailer. Gabe knew his role would be to program the robot to use its camera to track down opponents. He already started thinking about how to distinguish alliance robots from the enemy.

> "At the end of the autonomous period, human pilots step forward to take control. During this period of the game, they guide the robots as they attempt to launch moon rocks into the trailers of the opposing alliance. Meanwhile, human players also attempt to score moon rocks into passing trailers. With the low-friction playing surface, robots will slide easily, and high-speed collisions will be common."

On the screen, two pilots from each team, six per alliance, stepped up to the clear wall that separated them from the playing field and took the joystick controls. Their coaches stood behind them. Now under pilot command, three blue and three red robots raced about the field, chasing one another. Many shots at opponent trailers missed, leaving the field littered with balls. Robots got stuck or crashed into one another. At one point, a robot herded a bunch of balls into a low opening in the corner of the field, where a human player retrieved them and started shooting again.

> "During the last twenty seconds of the match, human players or robots can score bonus points by getting super cells into opposing trailers. Each scored moon rock is two points. And each super cell is worth fifteen points. Good luck."

"What have we gotten ourselves into?" Gabe asked.

"How are we *ever* going to go to Atlanta with this game?" another teammate said.

A D'Penguineers alumnus, who had come that morning to offer moral support, provided the opposite: "Oh man, this is going to be tough for you."

Once overcoming the initial shock, the students started sifting through the rule book. The warning against "polluting the well" was unnecessary—Gabe and the others were too busy just trying to get a handle on the game to even start thinking about design specifics.

The auditorium grew louder, everyone splintering off into cliques. A few students still focused on reading the rules. Others debated specific points, enthused about the game's similarity to basketball, or joked that their best bet was to build some kind of hovercraft. In between, they chatted about their winter break, moaned over an English paper due, picked through the breakfast spread, or shared earbuds from an iPod. By the time they broke for lunch, the auditorium was chaos.

This class of seniors had been thrown together to face this challenge, but they were not in any way a team. They hadn't greeted one another when they arrived. The kickoff presentation had temporarily united them, but now they were as disjointed as before, particularly without Amir there to guide them.

Gabe wondered, as did many of them that morning, if they would ever be able to work together. A year ago, Amir had even joked, "We may not do the competition with your class," and "I'd consider our season a success if we fielded a robot, any robot, without you guys killing each other first."

Although the academy drew many of the best and brightest from Dos Pueblos High School, this 2009 class had seemed bent on imploding from the beginning. When they first gathered as freshmen, there was more of the typical high school drama than in other years—jocks picking on the scrawniest nerds, kids ignoring the existence of others, the smartest teasing those who struggled. Kids just didn't treat one another well. Grudges developed. In sophomore year, a bloody fistfight broke out. By junior year, some students were failing to study, cheating on tests, tossing around racial slurs, and, in one instance, even dealing drugs in class right in front of Amir.

He tried to turn some of these students around, but he was forced to dismiss eight from the academy before the start of senior year. Plenty of others at Dos Pueblos were eager to join the academy, and Amir thought they deserved the chance. Eight new students came in as replacements. The troublemakers were gone, but the new students further fractured the class.

All thirty-one of the seniors were smart, but their differences defined them. Some came from wealthy families, others from households just scraping by. There were the quiet types who barely responded when addressed, and social kids sure to be on the short list for prom king and queen. There were varsity football and basketball players who wouldn't be caught dead at the math or science bowls that others pursued. There were kids who'd rather play on their computers than see the light of day, and gearheads who liked to fix cars. Others, like Gabe, were into the arts. Theater was his passion. A few others wanted to be classical musicians. Some arrived to class on skateboards and left school to surf until sunset.

At the start of their senior year, Amir had told them they needed to move past the bad blood and embrace their diversity. "This is our team. All of you are different, but your differences are what'll make us successful. . . . Here, there's a place for a programmer and someone who wants to work on metal. We don't have everybody here of the same temperament. In the real world, to run a business, you need that. At the end of this, you'll feel like a family."

For now, as everyone broke into packs to grab lunch before meeting Amir back at Dos Pueblos, family felt like a long way away.

———

"All right, so," Amir said, as if he had been carrying on a long internal monologue and was now giving voice to his thinking. He sat on the edge of a worktable in his Dos Pueblos classroom. A blank whiteboard stood off to his left.

Rising in their seats, his thirty-one seniors looked ready to burst out of the small steel-and-wood desks that were lined up in rows before Amir. Most of them had sat in this same classroom in these same desks to take Engineering Physics in their freshman year, then AP Physics in their junior year, both with Amir. Assembled again for their capstone project, they had outgrown the desks.

"This will be a crazy game, the hardest I've seen," Amir said, chopping the air with his hands. They were rarely still when he spoke. "It'll be chaotic, but let's start to figure it out. The most important thing to do right now is understand the game and come up with a strategy." He asked for volunteers to inventory the kit of parts to make sure everything was there. He wanted everyone else to break into teams of four to brainstorm ideas. "No idea is a stupid idea. One idea that might never work could lead someone to think of another that will."

Nobody in the room was more eager to get started than Chase Buchanan. In a way he had been preparing for this day since he was six years old, placing toys behind the rear wheels of his mother's car so that when she backed up and crushed them he could swoop in to examine their insides. The night before the kickoff, he had driven through a blizzard following a snowboarding trip to make it back in time for the *FIRST* unveiling. He was out of bed two hours before his alarm was set to go off.

The game animation had confused him as much as it had the others, but as the pilot of the team's robot (they'd had tryouts before winter break, navigating around trash cans with a previous year's robot), he was "stoked" that maneuvering the robot would be like driving on ice. Tough but fun.

At the start of the brainstorming session, Chase said, "The robot has to be able to score from any position." He argued that they needed a flexible shooter to compensate for how tough it would be to drive on the field.

Quickly though, questions about the rules diverted his group from any talk of what they should build. They reconvened to go through the rules yet again—and talk about potential strategies.

The session that followed resembled a bunch of people trying to put together a puzzle. They worked on one cluster of pieces only to get sidetracked by another cluster. If a piece was missing, they'd go through the rule book to figure out what it was. As the hours passed, the clusters came together to create a picture of how they wanted to play the game.

The Field. Two students laid down a whiteboard on a table and drew a scale model of the 27-by-54-foot rectangular "crater." Others created cardboard cutouts for six robots with trailers. They arranged them on the board to see how much open space would be left. "See, it's not crowded. No traffic jams," a student said before reiterating her argument that there would be plenty of room to maneuver on the field. However, the scale model also revealed that their trailers would be constantly under fire because of the positioning of the three human players on each alliance. Their robot would need to keep moving to avoid shots.

Robot Dimensions. Usually in *FIRST*, robots had to start the match with no part of their machine extending beyond a specified boxlike dimension. Once the match began, their robot could expand beyond this dimension with mechanical arms or other mechanisms. This year, with the slick floor, the rules were different. Robots could never extend beyond a 28-by-38-inch base and a height of 60 inches. The rules also required bumpers around most of their robot base. "We're talking about making robots that look like washing machines," Amir said.

Kit of Parts. The kit included sprockets, wheels, gears, bags of bolts, a camera, extension cables, a compact fan, an air compressor, yellow tubing, big motors and small, a lazy Susan, a joystick, master links for chains, bearings, a 12-volt battery, a simple chassis, and even a pair of gloves, among scores of other items. Except for the battery, motors, control system, and the slick wheels, which were required on the robot, teams could use as much or as little of the kit as they wanted. The kit material was only a starting point. *FIRST* allowed teams to spend thirty-five hundred dollars on raw materials to be fabricated into additional parts. In the end, there was a lot of room for ingenuity.

Trailers. They had a hexagonal base supported by two 6-inch wheels that were identical to the slick ones required on their robot. Given that the trailer had to be attached to the back of the robot, it would likely jackknife on turns. Short plastic pipes rose from the perimeter of the trailer base to form a circular basket that was roughly two feet wide. With the balls 9 inches in diameter, scoring shots did not mathematically look too hard. Still, they needed to build a trailer straightaway to see. Gabe said he felt confident that he could program the camera to track the trailer. Someone had the idea of tipping over the trailer to take away points, but the rules disallowed that strategy.

Game Pieces. They spent a good hour just correcting one another on the names: moon rocks, empty cells, and super cells. Turk kept tossing one of the balls from their kit up in the air, getting a feel for shooting.

With a possible 120 of the orange-and-purple moon rocks in play, each worth two points, these seemed to be the most important of the game pieces. Before the start of a match, officials gave each of the six teams twenty moon rocks to be shared between their robot and human player.

"People are going to be shooting balls," Amir said. "Robots are going to be shooting balls. There's going to be a lot of missed shots. Balls are going to be everywhere."

"We need to be able to pick them up from the floor," Chase said.

A few students, however, thought it was enough for their human player to load their robot with balls to score with, maybe tossing them into a big basket at the top of the robot. Amir and a couple of others countered that this would reduce the number of offensive threats on the alliance from six (three robots, three human players) to five. They needed to pick up from the field.

On the whiteboard at the front of the room, Amir had a student make three columns to list the abilities they thought their robot should have. From left to right, the column headings read: Absolutely Need; Want; Bells and Whistles. A ball-collection mechanism was first in the Absolutely Need column.

They would also need a way to store these moon rocks. Amir asked everyone to figure out how many balls they felt they could hold at any time given the maximum robot dimensions. Calculators flew from backpacks, and they figured out that they had room for about sixty balls. "How can you collect that many and have any time for anything else?" Amir asked. They decided they needed a mechanism to store a maximum of twenty to twenty-five balls at a time. It went up on the board in the Absolutely Need column as well.

As for the other game pieces, the team was almost unanimous

that they were more a distraction than anything else. Game officials gave each alliance four of the orange-and-blue empty cells before the match. Like moon rocks, they were worth two points each if shot into opposing trailers. But if an alliance exchanged an empty cell for one of the four green-and-purple super cells, and they shot this super cell into an opposing trailer, their alliance earned fifteen points. On first blush, this was a game changer.

The trouble with incorporating scoring super cells into their game strategy was twofold. In the first place, the exchange was complicated. An empty cell had to be put into play by the human player in the alliance who was positioned behind the wall at the center line of the field. An alliance robot then needed to deliver the empty cell to another human player positioned in the corner of the field. Only then could this corner player trade it for a super cell hanging on a hook by his side. Worse, super cells could not be shot until the final twenty seconds of the match.

The team figured they could score eight moon rocks, totaling sixteen points, in the same amount of time it would take for the exchange. For now they tabled doing anything about the super cells.

Autonomous. They tested how difficult it would be for a human player, or "payload specialist," to shoot over a 6½-foot wall (the height of the edge of the playing field) into a stationary trailer. Two students held a wooden plank to serve as a wall, and a trash can became a trailer. Turk shot. Pretty easy to score. During the fifteen-second autonomous period, their robot-trailer would be a sitting target. "Get the hell out of there" was the first priority. Using sensors, they should also be able to track down an opposing robot, collect balls from the floor, and shoot into opposing trailers. Gabe and the other programmers had their hands full. Add "multifunctional autonomous" to the Absolutely Need list.

How the Robot Scores. The students formulated two basic strategies. Chase championed the first: a shooter that propelled balls one at a time, preferably at a range of distances. Another student proposed that this shooter should be able to turn left and right to track a moving target like a battleship turret.

The second strategy involved a dumper where they could get up close and pour a bunch of moon rocks at once into an opponent's trailer. Many feared that opposing robots would be too hard to pin down to make this all-or-nothing score. Amir had a different reason not to pursue it: "I don't think a dump truck has enough going on. It's more important we build something cool than win the game. If we get out there, and a dump truck—where a gate opens and balls just roll out—wins, I don't know if I care."

They listed the shooter under Absolutely Need.

The Drive Train. This was the big X factor: how the robot would move about the field. They searched through the rule book for ways around the slick interplay between the regolith and kit wheels. Could they add wheels of a different material? No. Could they put tread on them? No. They brainstormed several other methods, but they were all illegal. There was no avoiding the low-friction field.

The team reminded Amir of his prediction before winter break that he was "99.9 percent sure" they would be using a six-wheel tank drive train that favored power over maneuverability. Now that they knew the game, tank was unlikely their best option. Maximum maneuverability was essential. They needed to construct a drive train where the robot could easily move in any direction, a daunting task that no D'Penguineers team had tried before. They added this to the Absolutely Need list as well.

After six hours, the puzzle started to come together. Amir was encouraged. The kids were engaged, and they had the sense of urgency that had been missing from their preseason meetings. It was clear that they understood that the strategy they came up with over the next few days would determine their future as a team. Amir also liked that they had ambitious plans for their robot.

Gabe joked that they didn't need three columns since Absolutely Need was the only one they used. The robot would drive in any direction, boast a highly functioning autonomous mode, collect and store a lot of balls, and shoot up to a distance of 12 feet in any direction. Building mechanisms to perform these functions would not be easy. But Amir knew that the more the team pushed themselves, the more they would take away from the experience.

At the end of the strategy session, Amir made two other additions to the first column. The first: "fit and finish." He wanted their robot to look professionally made—even beautiful—not a machine that looked cobbled together in the dark.

Second, he wanted to build two mechanically identical robots this season, something he had never done but that the best teams considered imperative. One would serve as a prototype for their designs and also as a practice robot for their team's drivers. This robot would be built first and used to work out all their design problems and mechanical issues. Only when this practice robot was operational would the team build their second: the competition robot. They would ship this machine by FedEx to their first regional competition by the build-season deadline, February 17.

"We're not going to pull any punches this year," Amir said. "I don't want to be conservative, questioning our ability to pull off a design choice or whether or not we have the time. It's either go big or go home, even though it's safer to keep it simple. This is your only year. You shouldn't have regrets that you didn't build something because it was too hard. We could fail, things might not work. I can't guarantee anything other than you'll be proud of the robot you build. But if it works, then we'll be good."

The students broke for a potluck dinner with their parents in the school cafeteria. Emily also attended. Amir showed the parents the game animation and said, "This chaos is why you won't be seeing your kids for six weeks." The parents chuckled, not sure if he was serious. He was. As he continued talking about the experience the kids would have, he started to trail off, exhausted from the long day. Emily was handing him note cards to prompt him. Finally he said to her, "Why don't you talk?" Emily jumped in to explain how fun the competitions were. She urged the parents to attend them, as well as to visit their kids during the building of the robot.

The parents left convinced, and the team stayed late to see if they were on the right track by choosing a shooter over a dumper. In a space cleared of tables, Chase and a few others held trash cans behind their backs and stutter-stepped around like robots with a trailer. With the single moon rock from their kit, Amir took jump shots from a range of distances, and except for a few errant throws that bounced off heads, he sunk them into the trash cans. Then he shuffled around like a robot, arms out, holding the ball in his hands, trying to passively drop it into the trash cans. It proved difficult to line up for the dump. They admittedly looked like idiots, but the exercise helped. They agreed a shooter was better than a dumper. Then they disbanded.

Amir and Emily returned to their 1950s tract house that they had stretched to buy on their teacher salaries. After a day that started with a 3:30 A.M. road trip, Amir was tired, but he couldn't stop thinking about the team's need to design a robot that could move in any direction. "This is key," he explained to Emily, who was both his sounding board and his emotional leveler. "We've no option, we have to do it. All the elite teams have experience with this kind of drive train, and they're going to do it and have an advantage."

Amir had never built such a drive train. Designing and fabricating one would take weeks of work, time they didn't have. It was also asking a lot of this band of students, but for their sake, and the academy, they needed to try. He spent several more hours that night bent over his computer, hand on his chin to keep his head upright, researching.

A few blocks away, Gabe cleared a heap of clothes and computer peripherals off his bed before crashing into it. The house seemed empty without his dad. Suffering from leukemia, he was now undergoing a bone-marrow transplant in Los Angeles. Gabe's mom was home that night, but more and more she would be in Los Angeles, leaving Gabe alone. With all his classes and homework and the two-hour drive each way to visit his father in the hospital every weekend, he already had a lot on his plate without the chief responsibility of programming his team's robot.

Turk stood in his bedroom, surrounded by Lakers posters and other fan memorabilia, including ticket stubs, bobble-heads, foam fingers, and even a mini Lakers basketball hoop hanging on his door. He lived in a small town house on the edge of the University of California, Santa Barbara (UCSB) campus. His father was from Lebanon, his mother from Honduras. There was nobody on the team who loved sports more than Turk. He looked at his heroes Kobe Bryant and Derek Fisher and wondered if he'd be chosen to be the human shooter over the one teammate who played varsity basketball. Turk dreamed of shooting moon rocks.

In a multimillion-dollar Mediterranean-style villa in the hills of Santa Barbara, Chase stayed awake long into the night, thinking about the robot they should build. Although his learning disabilities made reading and writing a struggle, he had a keen ability to see objects in three dimensions and put them together in his mind. He thought the base of the robot should have two independent axles to swerve around the field. The middle of the robot should be like a gun revolver, with balls in independent chambers to be pushed up to a shooter. This shooter should be able to rotate 360 degrees. Chase imagined driving the robot around the field, swerving around clusters of his opponents and pummeling them with shots. It would be beautiful.

Week One

43 DAYS UNTIL ROBOT SHIP

I said "nerds" and you flipped your head. If I said "bad-asses," then you should look up.

—CHASE BUCHANAN

The sun had yet to break on the horizon when Amir crawled out of bed on Monday, January 5. Winter break was over. He envied his still-sleeping wife, whose junior high school classes did not start until a couple of hours later than his at Dos Pueblos. Amir had only four classes to teach, but with robotics, he would not return home until 10 P.M. His foundation work, parent/student e-mails, lesson plans, and test corrections would keep him awake until at least 1 A.M. His capstone course, with its six-week build period and then competitions, would consume his coming months, so much so that he and Emily had planned the timing of her pregnancy around the season.

Amir quickly readied for work, knowing how much of his life was wasted in a typical morning routine. He had literally calculated it. The only place he took his time was the shower. It was a quiet place where he could think and run through what he had to do during the day. Some of his best revelations came to him there, though it was not without hazard. Occasionally, to his embarrassment, he would let slip to his robotics team where he had come up with his latest idea and suffer their teasing groans.

After showering, he dressed for class in slacks and a button-down shirt, grabbed his bag, and was out the front door. No breakfast, no coffee, nothing.

His cream-yellow stucco house, built on a postage-stamp-sized lot among rows of tract houses, was still a work-in-progress. Since they had bought the place, Amir had moved from room to room, tearing down the walls to their studs before rebuilding. He performed all the electrical and plumbing work as well. In "house-poor" Southern California, they were only able to afford a fixer-upper. Amir enjoyed the effort, learning cabinetry or how to put in a skylight or the kitchen sink, making sure everything was as they wanted. They planned on being there at least long enough to raise a family. Goleta was home, and as far as Amir was concerned, there was no other place in the world to live.

Set on a level bench of land between the Santa Ynez Mountains and the Pacific Ocean, Goleta had geography in its favor. In this town of thirty-three thousand, one could surf in the morning and hike in the steep, rugged mountains in the afternoon. Ranches and lemon groves stretched across the foothills. The weather added to the Garden of Eden effect: blue skies and 70 degrees Fahrenheit, with cool breezes at night providing natural air-conditioning.

Until the 1950s, Goleta was largely an agricultural community, far removed both culturally and economically from Santa Barbara 8 miles to the east. Then the building of UCSB and the arrival of a wave of aerospace companies, including Raytheon and Delco, transformed the town. Over the next two decades, the aerospace industry built thousands of houses for people working on the Apollo space program and military-defense systems. Everything was classified, and neighbors working for the same company had no idea what one another did during the day. At the start of the 1990s, most of these companies downsized, their facilities closing and getting sold off. Goleta survived with the start-up of several small technology and software companies, most founded by UCSB professors and their graduates. "Silicon Beach" they liked to call their town.

Less than five minutes after leaving his house, Amir drove into the Dos Pueblos parking lot. The high school was a collection of long, narrow, single-story buildings divided into classrooms. All the rooms opened onto covered outdoor walkways and courtyards with palm trees. Built during the influx of families to Goleta, the high school had experienced highs and lows, following the trends in the local economy. The year Amir had graduated, the school district wanted to close its doors. At that time, Dos Pueblos had the reputation of being a "stoner school" for the "hicks in the sticks." Students were known to leave the campus during lunchtime to get high in the nearby lemon groves. The school was down to roughly nine hundred students, and, as Amir said, was the district's "illegitimate stepchild." He and many other students protested to save the school.

Now Dos Pueblos enrolled twenty-four hundred students a year and boasted a reputation in the community for high academic achievement—in part because of the engineering academy. Its diverse student body included kids from across the economic spectrum. Some were the children of professors and software engineers, others of farmworkers, cooks, and janitors, more than a handful of whom were illegal immigrants. Classrooms looked like a study of America's proverbial melting pot.

In his first class of the day, Engineering Physics, Amir faced a roomful of giggly, doe-eyed freshmen. He lectured about temperature, explaining thermal conductivity by discussing the benefits of wearing a wet suit while surfing. "When you go bare-skinned into the ocean, your body is basically trying to heat up the whole ocean. Who do you think will win?" The freshmen answered: The ocean. "That's why you get hypothermia and die." Later, explaining that hot air rises, he had his students stand on their chairs to verify that the temperature near the ceiling was much warmer than near the classroom floor. Sometimes he asked questions to get the most obvious answer and would then respond, "Thank you. I wanted that answer because it's wrong." But he said it in a way that disarmed rather than humiliated.

Amir was a natural, yet for a long time, he had never even considered teaching.

Amir's road to becoming a teacher started with his father. Muhsin Abo-Shaeer came from a lower-middle-class family in Iraq and graduated from secondary school with the highest math and science test scores in his country. In 1959 his government sent him to Moscow to study physics, then to Columbia University in New York. After a youth spent in authoritarian Iraq, followed by life in totalitarian Soviet Russia, Muhsin found America a revelation. He earned his master's in theoretical physics, then followed up with a Ph.D. at UCSB, where he met Anita, a Pittsburgh native majoring in religious studies. For his postdoctoral work, he moved to Italy, and Anita joined him there.

Muhsin felt an obligation to return to Iraq, but he knew that doing so meant losing his ability to decide his future. His government would select his professorship, and any outside projects, for him. He delayed his return by accepting a professorship in São Paulo, Brazil. There, in March 1972, Anita gave birth to their first son, Amir. Soon afterward the family returned to California, Muhsin having decided that he wouldn't go back to the country of his birth.

In Goleta he had trouble finding a professorship because he was not a citizen and there was a glut of Ph.D.s looking for jobs.

He started translating technical documents with Anita, then took odd jobs like landscaping and building maintenance, to support his family.

The time away from theoretical physics gave him a chance to reflect. He had spent the previous twelve years in a world of functions and math, following a path his government had chosen for him. Now he was thinking independently, working when he pleased, and pursuing his own wide-ranging interests. He had an unstructured life and didn't plan on returning to any other.

He started a landscaping business where he was his own boss and could let his mind wander into politics, philosophy, cooking, poetry, or even physics as he worked. He repaid the Iraqi government for his education and rarely mentioned it to anybody. People in Goleta didn't know that the man mowing their lawns was a theoretical physicist. Many years later, a UCSB professor approached Amir and said that one of the great tragedies in his professional life was not being able to keep Muhsin in physics—he could have done great things.

Growing up, Amir felt like people looked down on his family because they never had much money. It also hurt Amir to see how people viewed his father around town. Since he was usually dressed in dirty, grass-stained pants from a day at work, people treated him like an uneducated immigrant. Amir wanted to scream, "My dad has a freaking Ph.D. in theoretical physics! He's a math genius!"

While at Dos Pueblos, Amir earned good grades, played drums in the marching band, and worked most evenings and weekends at the local grocery store, Vons. He and his brother, Jamil, both pursued undergraduate degrees in physics, Amir at UCSB, Jamil at UC Berkeley. Their father didn't push them in that direction, but it was almost as if they decided to pick up the path that he had abandoned. Jamil finished his Ph.D. at MIT. Amir veered off course.

By his junior year at UCSB, Amir felt a disconnect between the real world and the theory and math he was learning in class. He liked to see and feel how things worked. Solving integrals seemed impractical, and his professors never brought to life what was actually taking place in these equations. He finished his undergraduate degree in physics, still working nights at Vons to pay for college, then went on to graduate school in mechanical engineering. It was still theoretical, but stresses, forces, material failures— these he could visualize, and he was fascinated. He aced every class and won an internship at a local aerospace company, where he proofed out concepts for satellites. When he started, he was almost driven to tears at the little he knew about how engineers did their jobs, even with all his education.

After finishing his master's, he started on his Ph.D. He had four more classes and a thesis project to complete before obtaining his doctorate when he stopped. He realized he was pursuing it for the title and a certificate on his wall. He didn't need either for his self-worth. Later, he would see leaving the program as the moment he accepted the choices his father had made for his own life.

Amir took a position at Hendry Telephone Products. He hoped to return to the aerospace industry once they were hiring again. He had loved the research-and-development work there. A year and a half later he was still "building shelves" for electrical equipment. Aerospace was in a hiring freeze. At Hendry, Amir finished his week's work in two days, leaving him watching the clock and surrounded by engineers who had stopped trying to learn—and didn't need to for their job. He felt like Dilbert sitting in his cubicle. He could see the rest of his career as an engineer, designing different iterations of the same product, play out before him. He was sleepless, tense, suffering back pain.

"You have to quit," his soon-to-be wife, Emily, told him. They had met a few years before while they both worked at Vons. She didn't care about the money he was making. She had never seen him so unhappy. They started talking about what he should do. While studying for his master's, he had enjoyed his time as a teacher's assistant, turning around one student who went on to become class valedictorian. Amir had always liked telling people how things worked (he had partly won over Emily by explaining rainbows and showing her how he had crafted a magnifying glass he had given her from a bunch of spare parts). Teaching was a possibility. Emily was finishing her undergraduate degree at UCSB and was set on studying for her master's in education. They could get their degrees together.

Amir spent months on the decision. At one point, he decided to hedge his bets by attending night school for his graduate degree. Then one day at Hendry, a newly hired engineer was asking about thermoconductivity over lunch. Several colleagues offered explanations, but the engineer wasn't getting it. Amir held up a spoon and a cup of tea to show how the heat would flow. The engineer understood immediately, and one colleague said to Amir, "I'd never have thought to explain it that way." He looked at her and said, "That's why I want to become a teacher."

In the summer of 2001, Amir graduated from UCSB with his master's in education and started that fall at Dos Pueblos High School. For leaving a much more lucrative career in engineering to teach, a choice that his family and peers questioned, Amir credited his father: He had shown the courage to leave the path that was set out for him to pursue a different life. Amir had now done the same.

After lectures to his two AP Physics classes on how inefficient the internal combustion engine was—"If you like the environment, walk," he advised—Amir met with his robotics team in the build room. Besides their 9 A.M. class each day, they were scheduled to meet there Monday through Friday, from 3 to 9, and all day Saturday, until the February 17 deadline.

Forty-three days from now.

The build room resembled a shed. It was a 900-square-foot box with steel-panel walls and no insulation. Its only windows were set high along one wall, making it impossible to see more than a sliver of the sky. The school's architects had designed all the buildings at Dos Pueblos this way, working under the theory that it would prevent kids from being distracted by any activities going on outside. The room was on the edge of the Dos Pueblos campus, across from the recently renovated football stadium and the new aquatics complex. The school had given the team the raw space at the start of the school year. Over the previous months, his senior class had mounted the whiteboard, built the plywood workstations and shelves lining the room, wired the electrical, and hauled in a mill and a lathe. They placed these thousand-pound hunks of steel in a tiny office off the main room.

Some of their competitors were constructing their robots inside high-tech university labs or manufacturing plants boasting millions of dollars in equipment. This "machine shop" had no windows, and it couldn't fit more than six people without the risk of them bumping heads.

That first afternoon in the build room, Amir wanted the students to break into five groups: programming, electrical, and three mechanical crews to focus on the drive train, intake/ball-storage mechanisms, and the shooter. He assigned John Kim, who served as a kind of teacher's aide, to form these groups.

It seemed a simple order, but its execution was anything but. John found programming easiest. There were only four students, Gabe foremost among them, who knew enough to be in this group. Selections for the other groups were a nightmare. Everyone seemed to have several people they refused to work with. The matrix that John created to avoid these pairings taxed even his highly skilled math abilities. Such was their team at that point in the season.

"It's engineering, not inventing," Amir declared after John set the groups, advising them to check out robotics books and Chief Delphi.com, the website for all things *FIRST*. "We need to keep looking at other ideas and find out what works best. Then make things better."

Throughout the first week, the build room was a scatter of different activities overseen by Amir and the small crew of mentors he had recruited from the local area to help the team. They included Stan Reifel, a fifty-year-old Internet entrepreneur; Danny Lang, a twenty-eight-year-old electrical engineer at Raytheon; and David and Juliana Boy, a husband-and-wife pair studying for their graduate degrees in engineering at UCSB.

Guided by David and Juliana, the shooter team made fast progress. Their first shooter prototype operated like a baseball pitching machine. Balls fed between two spinning wheels and shot out the other side. It was straightforward and was constructed in less than half a day, chiefly by Daniel Huthsing, a Boy Scout with an irrepressible need to please, and Yidi Wang, a tall and bubbly concert pianist who was one of the academy's eight new members.

When they powered up the drills that they had rigged to spin the wheels, Daniel was standing right in front of the shooter and got pummeled by a moon rock. The next time, he ducked below the shooter. The wheels spun, and over the noise of the drills, he squinted as if about to get hit again and yelled, "Fire!" Yidi fed the moon rock into the prototype. It compressed between the wheels and shot out the other side straight into the trash can they had set up as a basket. Score.

They moved the basket back, increased the speed on the drill, and tried again. Miss.

Different speed. Miss.

"Ahh, snap," Yidi said, a phrase she always used to express something bad. "Oh, snap" was good.

Daniel moved the trash can to its first position, a foot away from the shooter. "It makes all of them from this distance."

"I'd hope so," deadpanned another student.

They attempted a range of drill speeds, but the prototype was inconsistent with its shots. Further, the whole thing was too bulky to sit on the top of the robot and rotate like a battleship turret. They moved on to their next idea.

Made out of plywood and bendable hard plastic, this shooter looked like the curved hood of a snowblower. Balls would start within a short column below the hood. When pushed upward, they would meet a treaded wheel that spun at the base of the hood's opening. This wheel would force the balls to travel the curved path of the hood until being propelled toward the target.

The three girls on the shooter crew built the hood. The three guys built its frame. Everyone gathered for the first test run. Daniel powered the drill connected to the end of the rod that spun the wheel. The whole contraption, held together by some screws, C-clamps, and a bottle of glue, shook on its plywood mount as the *whirr* of the drill filled the room. Another student stood several steps back ready to catch the shot. Yidi eased a ball up to the wheel. The ball begrudgingly rolled up the hood and, thanks to gravity, left the opening.

"Great, we have a drooler," Yidi said.

There was too much space between the wheel and the curved

hood to keep the ball compressed and moving consistently forward. "Not enough squeeze," several students said. Yidi and Daniel spent a rushed hour adding foam inserts to the hood to narrow the balls' path. By the time they finished, Yidi had glued a pair of her fingers together.

Test two. Yidi fed another ball up to the spinning wheel, and it launched from the hood with a *swoosh*. Amir clapped. "Oh, snap," Yidi said before dancing a jig. Others in the shooter group hugged. On the right track, they continued to refine the prototype, adding an adjustable cover over the fixed hood to change the angle of their shots.

Those working on the intake roller, the mechanism that Hoovered balls from the field into the base of the robot, made even quicker strides. Stuart Sherwin, whose thick curly mane of brown hair flowed down past his shoulders and hid his eyes, built a plywood U and laid it flat on the floor to act as a kind of corral. At its open end, he mounted a long PVC pipe parallel to and 8 inches above the floor. They covered the pipe with the same rubber tread used on the shooter, a grippy material often employed on conveyor belts. A power drill spun the PVC pipe as Stuart rolled a ball toward it. The contraption squeezed the 9-inch-diameter ball between the floor and spinning PVC pipe and sucked it into the corral. Easy.

Despite this progress, they still did not know how the robot would drive, where the moon rocks would be stored, and how they would reach the shooter. Amir bounced from one group of students prototyping ideas to the next, but he was primarily focused on fixing up the machine shop. They would fabricate most of their robot in the shop, so it had to be ready for operation once the team settled on a design. In addition, the time in the shop gave students more opportunity to generate ideas and learn basic construction practices. As mentor Stan joked, some seniors didn't know which end of a hammer to hold when they first walked into the build room.

That first week, with Amir guiding them, the students cleaned off the rust and recalibrated the lathe, drill press, and milling machine. They installed the lines for compressed air. They built a welding table. They made aluminum wall mounts for all the drill bits and tools they would need ready at hand.

Chase spent so much time standing at the mill, calibrating it to make cuts to thousandths of an inch, that his feet swelled up. He found it hard to remain in one place hour after hour, repeating the same steps again and again, in a room as cramped as an elevator. Every now and then he skateboarded around the campus to relax. Midweek he traded his flat-soled sneakers for a pair of shoes with Velcro straps he bought at a thrift store. Made for the elderly, the shoes looked ridiculous, but their arch support and padded soles relieved some of his discomfort. He branded them his "machining shoes" and tricked his mind into thinking they helped him concentrate. Steadily the machine shop came together.

Between prototyping and working in the machine shop, students played around with shooting moon rocks into a trailer they had built to the competition specifications. Anthony Turk spent more time than anybody hurling balls into the trailer, already worried that Bryan Heller, the team's only varsity basketball player, would be their human shooter. The fact that Amir would often say, "When Bryan is shooting..." when talking about this aspect of the game only contributed to Turk's anxiety.

"Under pressure, Shaeer," Turk said, sinking another shot, "you can count on me. I'm no underdog. I'll come out on top."

"I can see it now," Amir said with a chuckle. "Students going over to see Turk at his house. He'll say, 'Nope, can't come out. I'm studying.' Really, he'll be practicing, his mom dragging little trailers around the house to help." This was classic Amir, using humor to make his point. He knew that Turk wanted to be the human shooter, but he'd have to earn it.

Along the wall nearest to the machine shop, Gabe and the other programmers lived in their own little world. Amir didn't know much about writing software for the robot, so he left them to work it out. Their first goal: Program the camera to identify the bright pink-and-green band at the top of the center post in each trailer. They would use this "vision target" to track their opponents. Although the programmers worked separately from the others, they seemed to keep a keen ear out for everything said in the build room.

"Nerds," Chase said one afternoon while exiting the machine shop.

One of the team's programmers, Nick Vaughn, turned in his chair and glared at Chase. His face darkened. "Who are you talking about?"

The tension among the students rarely surfaced but was evident in how little interaction there was among the self-selected groups. Now though, the build room turned quiet.

"What? I didn't say 'nerds' and point at you," said Chase, who considered himself a nerd anyway. "I said 'nerds' and you flipped your head. If I said 'badasses,' then you should look up, but not 'nerds.'"

His words broke the tension—at least for the moment. Everybody laughed, and Nick smiled uneasily.

Throughout the first week, students in the drive-train crew spent most of their time deciding if they needed to design the "über-complex" system in which the wheels could move in any direction. Although the team figured maneuverability on this low-friction field was the key to the game, one didn't need to build a rocket ship to fly into the clouds if a hot-air balloon would do. It was worth a few tests.

Both in class and in the build room, Amir ran through the theory of friction. "We don't want to accept the fate of the field," he said, explaining that they had to find a way to overcome the slippery surface. "The drive train, it's a crazy physics problem. It's the one thing you can mathematically characterize. Everything else is a little physics, a little gut, a little trial and error." The students, half of whom had aced their AP Physics exam, thanks to Amir, didn't need the refresher course on friction, but he gave it to them anyway.

"With friction, you care about two things: first, the nature of the two surfaces interacting—there's more friction resisting movement with a rubber tire on pavement than on ice—and second, how much force is pushing those two surfaces together. Basically, this is weight. It's easier to slide a four-legged table across the floor if there's not an elephant on top of it. Most people say more surface area, more stuff touching the ground, more friction. Not true. If you have more surface area, you're distributing weight over more area. Let's say we added twenty wheels to the robot; each wheel would carry one-twentieth of the weight, so there would be less force of friction for any wheel on the surface. That's the common thing people screw up. Remember, we don't need six million wheels. We only need the number of wheels necessary to balance the robot—the fewer the better.

"We also have to deal with this concept of the coefficient of friction," Amir continued. "Remember, the coefficient is a measure of how much friction there is between two surfaces. The bigger the coefficient, the more friction."

Rubber-treaded wheels on the typical *FIRST* field of carpet had a coefficient of roughly 1.2, Amir explained, meaning 180 pounds of force would be required to slide a 150-pound robot across the field. According to this year's *FIRST* rule book, the frictional coefficient between the regolith and the kit wheels was 0.1. That meant it would take a mere 15 pounds of force to slide a 150-pound robot across the field. This was twelve times less traction than on a typical field, and the reason that at the Manchester kickoff, Dean Kamen easily pushed the boxlike sample robot across the field with his foot.

The team would want to power their drive train so that the force the wheels applied to the ground from a standing start would not exceed 15 pounds. Any force greater than that, and the wheels would slip. Amir had them imagine a sprinter. For the quickest start, she would want to push off against her back leg to the maximum amount she could without her foot slipping on the track. The same with their robot. Below 15 pounds of force, the wheel would rotate on the regolith, pushing against it but not slipping, enabling the robot to go forward.

Then life got complicated. There is more friction between two objects when they are stationary (static) than when they're already sliding (kinetic). A quick reference to the rule book showed that if the wheels were already slipping along the regolith, the friction coefficient would be reduced by 25 percent. In other words, they would have even less traction once the robot was actually in motion, making it even easier for the robot's wheels to spin out.

Imagine trying to move a refrigerator up against a wall, Amir told them. You have to push hard to get it to move, but then you ease back once it starts sliding because otherwise you'll push it through the wall. Once the refrigerator is sliding, the kinetic friction between floor and refrigerator is much less than the static friction before it was first budged.

The theory aside, Amir wanted the team to see how this played out practically before they made a final decision on the drive-train design. He tapped Chase, who had pushed hard to work on this part of the robot, to lead the tests. Their initial task was to see if the friction coefficients in the rule book matched the interaction between their kit wheels and the regolith sheets they had bought at a hardware store.

Following *FIRST*'s instructions, Chase constructed the simple chassis provided in the kit. It was basically an erector set where one bolted the individual parts together. Viewed from the top, the chassis looked like two long narrow rectangles placed side by side and tied together with two cross members so that it made a box shape.

Without much trouble, Chase put together the left rectangle and set two wheels on its front and back end. The other members of his drive-train group, including two girls, Saher Hamdani and Lisa Nakashima, whose pixieish appearance made them look as if they'd gotten lost on the way to the cheerleading squad, dove right in and mimicked what Chase did on the right side. They then slid bolts through the wheels to prevent them from rotating.

Now they had a problem: How would they simulate the 150-pound weight of the robot for their first test? The chassis they had built weighed 50 pounds. The drive-train group looked around the room and spied Angie Dai putting together a basic electrical board. They weighed her on the industrial scale: 100 pounds even. "You'll get a good workout," said Saher to help convince Angie that perching herself like a cat on the chassis would be okay.

Angie stretched her slight body across the frame. The daughter of two Chinese math professors, she scored 2,380 on her SATs, one question from perfect, and already had an offer to attend MIT in her back pocket. Now she was just a weight source.

Chase pulled the frame across the regolith with a spring scale, taking a reading on how much force was needed to first budge the robot, then another on how much to keep it sliding. Not much force at all for either. Angie proved even more useful when she ran through some calculations to derive the friction coefficients. After they took a few readings, their numbers approximately matched the ones provided in the rule book.

Next, Chase and his crew put together one of the simplest, most popular *FIRST* drive trains for the chassis—a tank drive— to create an operable robot base. Chase locked all four wheels straight forward, preventing them from turning. He then linked together the front and back wheels on each side of the chassis with a chain and sprockets. A single motor powered each side.

This type of drive train operated like a military tank, but instead of treads it used wheels. In order for it to go forward or backward, the wheels on both sides of the robot spun in the same direction at the same speed. To turn left, or more accurately, skid left, the wheels on the right side of the robot ran faster than the wheels on the left. To turn right, the opposite was true. To pivot around on a fixed point, the wheels on either side moved in opposite directions at the same speed.

If they could get away with using a four- or six-wheel tank drive on the regolith, they would have a lot more time to develop other robot mechanisms. On Friday afternoon, January 9, they placed several sheets of regolith in a single line to test their prototype robot base. Chase was more pumped than anyone because this was the first system he had built. Once again, Angie climbed onto the frame.

"Hold on," advised some of her teammates.

"Don't die," said others.

Gabe had programmed some basic controls to operate the robot base with a pair of joysticks. Chase was driving. When he pushed the joysticks full forward, the robot base skittered across the regolith like an insect. Then, with a sharp screech, the wheels burned out.

On the following test, Chase eased the joysticks forward. The wheels rotated without slipping, and the robot base gathered enough speed that Angie white-knuckled the frame. Given how delicately Chase had to ramp up the power to the motors, they clearly needed some kind of traction control to keep from burning out.

Next they wanted to see how the tank drive turned on the regolith. Chase drove the robot base forward a short distance before pushing the right joystick farther forward than the left, causing the right set of wheels to spin faster. The base turned left, almost fishtailing out of control.

"It's a lot better than I thought," one student said.

Amir withheld comment. Chase tried to recover from the turn, but the wheels slipped and the screech returned.

Finally, Chase brought the robot base under control. Amir had seen enough. He had hoped these tests would somehow prove that the tank would be okay, but it was obvious now that it would not. He kept his conclusion to himself, however, wanting the students to come to it on their own.

The next day they continued with their tests. Getting some of the kids in early on their first Saturday was tough. When Amir asked the team where Stuart Sherwin was, his closest friend said, "Uh, he likes to sleep in on weekends." Amir looked at the clock, which read 10:30 A.M., and smiled. "Call him anyway."

Now they needed to see how the tank drive would perform with the trailer attached. When the robot base was moving at a good speed, Chase attempted a turn, and the tank drive performed as poorly as the day before without the trailer.

Then he tried to turn the robot base from a standing start. It jackknifed with the trailer almost immediately. Maneuvering out of the jackknife was slow, ugly, and involved a lot of slipping wheels. The second time Chase ran the test, he couldn't escape the jackknife at all.

"Disaster," one student said.

"Terrible," said another.

Chase just stared at the robot base. "The trailer makes it so much worse."

"Are we happy with driving like this?" Amir asked. As usual on Saturdays, he was unshaven and in shorts. "Is this what we'll show up with?"

Several students shook their heads. Amir did as well.

"Cool. We get to make the gnarly drive train," Chase said.

On Saturday night, the end of their first week, the team met in the cafeteria for a tri-tip beef and garlic bread dinner hosted by one of the parents. They had dinner together every day, one of the ways Amir hoped to build team unity. As always before the meal, the students said to the parents, "On behalf of the Dos Pueblos Engineering Academy, we thank you for bringing this food." The students clapped, then rushed the line.

This was the one time during the day when they could unwind a bit. Even Amir, who seemed to have a clock in his head to thwart wasted time, joked around with his mentors without feeling the need to rush back to work. When one of them wondered aloud when dinner should end, Amir put him at ease.

"If I'm worried about eating in week one, we're screwed," Amir said. "Check back in week six. I'll be in a daze walking around with an empty plate in my hand."

Near the end of dinner, there was a conspiratorial burst of

laughter at the far end of the tables. In an instant, the kids' exhaustion had cleared away with a new experiment.

Kevin Wojcik, who along with John Kim had joined Amir on the trip down to Los Angeles on kickoff day, rubbed the seat of his pants back and forth on a plastic chair. Chase stuck his finger into a puddle of water that had spilled on the table. After generating enough static electricity, Kevin put his finger into the puddle as well.

Zap.

Chase jumped up. The kids laughed, and more of them gathered around. One poured a line of water onto the table. Kevin shifted back and forth again, placed his finger in line.

Zap.

The air was so dry they could see the spark. The kids grew inventive. One poured a little salt into the water. Another extended the trail of water. John, who looked like he hadn't slept in weeks, dumped half his water bottle onto the table, drawing a roar of laughter from the team. Then he stuck his finger in the center and announced, "C'mon! I'm ready! It's an event from the Brain Olympics!"

Zap.

Now a "team-bonding" session, Gabe declared. Four kids placed their fingers in the water and shouted in unison, "Team 1717!" Kevin shifted in his chair again, but there was no shock. John and Kevin locked index fingers as they shifted in their seats together. When ready, they rose from their chairs, moved toward the water, and—

ZAP.

Kevin fell back in his chair, laughing wildly. John backed away, a little stunned. "That one hurt!"

Kevin shook his finger. "Experiment o-v-e-r."

Looking on from across the room, Amir laughed. It was nice to see that they were using what he had taught them, even for Beavis and Butt-Head–like behavior. It was nicer still to see everyone gathered in a circle, joking around with one another, instead of divided into their usual cliques. For the first time, Amir thought this class of students with its "bad spirit" might have a chance at coming together as a team.

Week Two

36 DAYS UNTIL ROBOT SHIP

What would kill me is if we get the robot out there and it falls over with smoke coming out of its side.

—STAN REIFEL

sking a team of rookies to design, fabricate, program, and assemble a sophisticated robot in six weeks was no small matter. But Amir also expected his students to be ambassadors and, in a way, volunteer staff for the engineering academy. They prepared presentations for potential donors to the academy's fund-raising campaign. They produced the website and other publicity material. They pitched media. They designed their own uniforms and logos to give the team a slick look at competitions. They tracked expenses and kept the books for the team. Every team member even had a business card.

On Monday, January 12, the team would be putting those ambassadorial skills to work in courting a special visitor, Congresswoman Lois Capps. While practicing how to operate the academy's 2008 robot for the presentation, Gabe hit a button on the joystick. The forklift arms launched upward much faster than he had expected and broke through the tiled ceiling in the build room. He froze, staring upward. It was 3:15 P.M., moments from the congresswoman's arrival. A local television news crew and several journalists would be trailing her on the tour.

"I was using the right one, but something—" Gabe stopped himself.

Amir examined the broken ceiling, then glanced at his now speechless programmer. "You know what you did wrong, don't you?"

"Yes."

"Okay, let's not do that again," Amir said. Instead of his usual frenetic movements about the build room, he sat still in a chair.

Their imminent visit with Lois Capps, who had prompted this meeting on word of the academy's success, reminded Amir once again that he was depending on this team's season and their robot to showcase the work he was doing at Dos Pueblos.

At three-thirty, with no sign of the congresswoman, a shy and bookish Isabelle D'Arcy, a member of the electrical team, was overcome with nerves. She asked Amir if she could pass off her portion of the presentation on the academy's public relations to another student and only talk about the logos on their flight suits, the team uniform. "I don't want to blow it," Isabelle said, the retainer in her mouth making her a little difficult to understand.

"Well, if you do, you'll blow it big, in front of Congress," Amir said before reassuring her. "You'll do fine."

Amir moved some more girls to the front of the two concentric semicircles in which the team had arranged themselves. He also asked the students slurping at their Slurpees to find the nearest trash can. Colin Ristig, a surfer with a sweep of thick blond hair who always knew when some levity was needed, asked, "How do you think Capps gets around? By helicopter or teleport?"

The team voted teleport.

Finally, Lois Capps walked into the build room, cameras rolling. A six-term congresswoman, she was a 6-foot beanpole in a tan pants suit, the blond hair drawn around her face disguising her seventy-one years. She introduced herself to the team, saying that she was a former school nurse.

After she took a seat, Amir stood to give an overview of his academy. He was not a natural salesman, and like Dean Kamen, he tended to get lost down tangents, feeling the need to describe his program's every nuance, but there was no doubting his conviction. Nearing the end of his introduction, he said, "My goal's not to push students to do engineering but to expose them to an exciting approach to science. If they're interested, they can take it from there." He concluded by telling Capps that the students leave the academy knowing much more than how to create a robot.

Alyssa Ogi, who had the beaming smile of a pep-squad leader, said, "I'm not planning on having engineering as my major, but I'm building skills that you can use outside that." She described how each student worked not only on a part of the robot but also on an aspect of the academy. "We do the accounting, human resources, graphic design, writing, presentations, record keeping, database programming, and public relations."

Isabelle jumped in on her cue, describing how they promoted the team's activities to the media and recruited sponsors. "We even design the logos for our corporate sponsors," she said.

Others followed. They spoke of how each student learned a specialized skill to build the robot, like using computer-aided drafting (CAD) software to design parts, and then taught the others this skill. They highlighted that the team was almost 50 percent female, a fact that challenged many people's assumption that girls wouldn't be interested in this kind of activity.

Gabe explained that unlike any other team in *FIRST*, they were all rookies, all seniors, and this was an actual class where they would receive individual grades. "Most teams are after-school clubs with kids from all grades. If they graduate eight kids a year, they only lose eight kids. Our team is new every year, and Mr. Shaeer has to teach everything all over again."

"The academy's about analytical thinking, problem solving, and working with the team," Amir said, punctuating each word. "I think that we, as a country, need to improve our science and engineering, but I'm more of an advocate for *all* of our education to be structured like this—be it a business academy, engineering academy, et cetera. That's the whole point. We're so focused on standardized tests, benchmarks and things, that while they may have some meaning, they're not the end-all be-all.

"My stump speech is this: Students are on the Internet these days, and information's free. If we teachers don't move beyond just being fountains of information, and instead focus on experiences, we're losing touch with what students need. And the students *know* it. They're sitting in class thinking, 'I can look up this stuff on Wikipedia right now.' They can't look *this* experience up on Wikipedia."

"Let me ask you then," Capps said. "I understand what you're saying about standardized tests, but that's the game being played now in our country, so how does this fit into that?"

Amir began to answer but hesitated, as if debating how far he should go in front of the congresswoman and her cameras. "I've engineered this program within what I think is an asinine educational system." He smiled, as if to cushion the statement. His students grinned. "I've looked at it and said, 'Here it is. I can't change this system.'"

"Right," Capps said.

"It's like wading through peanut butter, you know what I mean?"

"Right." She nodded, affirming again like the flock to the minister.

"So, in parallel, I'm going to create a program that works within the system, demonstrate that it works, and change the system from the inside out."

"Yes," Capps said.

After a showcase of the 2008 robot by a now overly careful Gabe, Capps concluded, "I'm going to be looking to you all for guidance as we push this further nationally."

Before the congresswoman had finished, John started a slow clap in the back of the room. Nobody else clapped, and he stopped, a sheepish look on his face. As Capps made her final point, a few students chuckled at John for being "the guy who clapped too early."

After Amir and his students thanked the congresswoman and did a few television interviews (the kids scrambling to get in the camera's frame), it was time to get back to work. In the doorway, Amir stretched backward while placing his hands over his eyes. The presentation had gone well, very well, and it would be great for the academy to have someone in Washington, D.C., who might spread the message of what they were doing here in distant Goleta. But for now, Amir announced, "God, I just want to build robots."

Soon after in the machine shop, Chase Buchanan drilled three large holes into a piece of plywood. When he finished, he inspected the cuts for the three-wheel prototype robot base he was building. He removed his safety glasses and dangled them from the front pocket of his faded jeans, his white T-shirt hanging loosely on his thin frame. Now and again he ran his hand along the side of his face as if to see if he could feel the few straggles of facial hair that had grown in the week since he last shaved.

Chase was a handsome kid who dressed kind of grunge hip. He lived in a "sweet crib" that his classmates envied. He skateboarded and rode single-speed, fixed-gear bikes with no brakes down steep hills. He listened more than he spoke, rarely trying to impress, and he treated everyone on the team the same, not seeming to buy into one group or the other. What defined Chase was his supreme ease. He had the air of someone who was comfortable with who he was and had found where he needed to be. Of all the kids on the team, probably nobody else fit better within the academy.

Chase came from a family of inventors and engineers. As a young man, his grandfather, who had a Ph.D. in systems engineering, went to an interview at a military defense contractor, and before they even had a chance to offer him a job, he had already made a significant correction to one of their most important projects. He went on to help develop the Stinger missile and other weapons systems that would become the pride of the United States military. Chase's father was an endodontist. His own inventor streak had made him a fortune when he came up with the design of a single-barbed file used in root canals that had previously required eighteen separate files.

For a homework assignment titled "My Future Career," an eight-year-old Chase had drawn three interlocking gears in pencil and written, "I think my future will be in inventing. I wouldn't be in any particular classification. It would just be anything I could get my hands on. I have wanted to be an inventor ever since I could say my name." Growing up, Chase built pinewood derby cars and an elaborate tree house with his father instead of throwing a ball around the yard. He disassembled McDonald's toys to see how they worked. He pried motors out of radio-controlled cars and put them in balsa-wood cutouts to launch across his pool. He constantly drew things he wanted to build, boats with water cannons, a military ambulance with a rocket launcher. He even rendered exploded views of these creations to show how the individual parts came together.

Over time, his acute ability with spatial relationships seemed to strengthen. During a summer internship at UCSB, Chase amazed his supervising professor with how quickly he assembled experiments. He skipped steps that most people would have to take because he saw in his head how everything worked together.

However, when Chase tried to read words on a page, they jumbled. But dyslexia wasn't his only challenge. Doctors had also also diagnosed him with dysgraphia, which mixed up his words when he tried to write, as well as attention deficit hyperactivity disorder. As Chase said, "I'm terrible at reading and terrible at writing. And because of my ADHD, I can't pay attention to what I'm *not* reading and writing." At one point he took medication to counter some of these "crossed wires," but he felt the drugs killed his creativity, so he stopped taking them.

Chase attended an expensive Santa Barbara private school from kindergarten through eighth grade. Although he did well in math and science, his other classes challenged him. No matter the class, he found it difficult to sit through his lectures and some days brought in LEGOs, assembling them under his desk to keep from going mad with the need to move.

While his close friends went to Santa Barbara High School, Chase decided to go to Dos Pueblos because of the engineering academy. Although entrance to the academy that year was based on a lottery, his mother called, harassed, angled, and did everything she could do to get her son accepted to the program, even approaching Amir at the funeral of a mutual acquaintance. In the end, luck won Chase one of the limited number of spots. In his first year at the academy, where he excelled at the Rube Goldberg projects, he came to understand his learning disabilities as part of "what makes me *me*."

He looked at the academy's capstone course as his big test, the way to prove that his future lay in inventing and engineering.

Since high school began, he had been counting down the days for the robotics season to begin. He also came into it thinking he had the most experience designing and constructing things of anybody on the team. He wanted to make sure his ideas, whether game strategy or mechanism design, were at the forefront. In particular, he wanted the drive train to be his baby.

With a tank drive now out, Chase focused on how to build a system with maximum maneuverability. Like Amir, Chase had scoured the Web researching various robot drive trains. They had names like crab and swerve drives, but the differences between them seemed small, and people used the names so interchangeably that it was hard to decipher which was which. Their basic premise was the same, however: They had wheels that were independently driven and could turn 360 degrees.

Picture a typical office chair with swivel wheels. Then imagine that instead of pushing off the floor with your feet to move the chair in any direction you wanted, a joystick controlled independent motors attached to each wheel. If you want to move left, you push the joystick in that direction. Like little soldiers, the wheels would pivot left and start rotating together at the same speed, moving the chair where you want to go. That is essentially what the team was trying to build.

They didn't have any step-by-step instructions with exploded views of how all the parts came together to build such a drive train, only a few pictures and rough descriptions from other teams of how their own had worked. Some operated with three wheels, others four, but they all looked complicated. Chase understood that there needed to be one set of motors to drive the wheels and a second set to steer them. Both sets needed independent transmissions.

An electric motor provides a rotational force called torque. The more torque a motor provides, the harder it is to stop its output shaft from spinning. Of the several types of motors contained in the kit of parts, the black cylindrical CIMs (the acronym for its manufacturer) were the most powerful, able to spin at close to 5,300 revolutions per minute, as long as there is no resistance. But as resistance is added, the motor slows. If the motor is connected directly to the wheels, the weight of the robot would create so much resistance that the motor would stop and the wheels would not turn at all.

The wheels needed more torque than the motors could provide. A transmission would solve that issue by converting the high speed of the motor's output shaft into more torque on the wheel. In exchange, the wheel would spin much more slowly than the motor's shaft. More torque, less speed—that was the trade-off.

Think of riding a ten-speed bicycle. On a flat surface you don't need to provide high torque to the wheels to make the bike move forward. When you encounter a steep hill, however, the wheels need more torque to allow you to make the climb. In this situation, you, just like the motor, are too weak to provide the necessary torque to the wheels. By switching gears, you can pedal faster and the wheels spin more slowly but with more torque. Different-sized sprockets connected by chains (a simple transmission) exchange your high pedal speed for greater torque to the wheels. This allows you to continue traveling up the hill but at a much slower speed than on the flat surface.

For their robot to best maneuver on the regolith field, Chase and the rest of the team would have to build a transmission that would provide the optimal relationship between torque and speed. All of these motors, transmissions, wheels, and axles would need to fit within the chassis. There would be scores of custom parts to fabricate in the machine shop. Sensors would be needed to determine which direction the wheels pointed and how fast they rotated. The electrical team would have to wire each motor and sensor, and the programmers would have to figure out how to evaluate what the sensors were saying and utilize that information to determine where the robot should go. It was a tall order, but unless they wanted to spend every match sliding around, unable to evade other robots, getting pinned against the field walls and chasing after balls, they needed an advanced drive train.

For now, Chase only had to see if a three-wheel drive train would be stable enough on the field. Four wheels would be more stable, but three wheels would need fewer motors and promised to be less complicated to manufacture.

With the help of his close friend Colin Ristig, who restored cars with his dad on weekends, Chase finished setting three wheels in the plywood prototype base. Now they needed to see how easily the base would tip over. They recruited Saher Hamdani to stand on the base to replicate the right robot weight, but she was too light on her own. They then loaded her arms up with a toolbox and a couple of batteries. She was laughing so hard that she almost tipped over.

"What's going on?" Amir asked. "This isn't exactly the pinnacle of safety, so can you lower the humor level like five notches?"

When Saher's arms gave out, Amir stepped onto the base and said to Chase, "Try to tip me. But don't get too aggressive, I'm old."

Chase failed to push him over, but mentor Stan Reifel, advising in his usual straightforward way, said, "You're absorbing the blow each time, Amir. It's a useless test."

Stan, a jack-of-all-trades kind of engineer who was Amir's second-in-command, suggested piling a bunch of large storage totes on top of the prototype base. While Chase ran around the room collecting and filling the bins, Amir whispered something to Kevin. A wide grin stretched across the programmer's face as he walked away saying, "This is going to be awesome."

"What's up?" Gabe asked.

"We're going to ram last year's robot into it."

"Sweet."

Amir and Chase positioned the prototype onto a sheet of regolith. Kevin and Gabe placed the 2008 robot, which looked like a compact forklift, 15 feet away. Once it was set, Amir dropped his arm as if signaling the start of a drag race. The robot sped forward and rammed into the prototype base. *Smack.*

But the prototype remained upright.

They repeated the exercise, increasing the speed of the ramming robot each time. The base Chase had built took the beating like a champ, not tipping. Later that night though, Amir told Chase that he wanted to put off any decisions on the drive train, including whether it would have three or four wheels, until they figured out how they would both store balls and deliver them from the intake mechanism to the shooter.

"For those of you sitting around and pondering, you don't get to sit around and ponder. Engineers working for companies get to do that. We've a robot to build," Amir announced to the team, who had been milling about, undirected, since they had finished their Chinese dinner on Tuesday night. "You need a blue-collar attitude. A better work ethic."

It was thirty-five days until the ship date of their competition robot.

Amir returned to the machine shop, where he was trying to fix the drill press. A moment later he came back out with a metal tube and handed it to Turk. "I need you to measure the diameter of this with your calipers. I'm going to base everything on your measurement, so you better get it right."

Turk got a digital caliper and measured the tube between its jaws. "It's .53 inches."

Amir looked at Turk, a slight smile on his face. "Really?"

"Here we go." Turk rolled his eyes, knowing he got it wrong. He stuffed his hands into his jeans and started rocking foot to foot, waiting for one of his teacher's impromptu lectures.

"Just looking at it, do you think it's a half an inch?" Amir asked.

"Yeahhhhh, I guess."

Several students around him chuckled. Turk always took these exchanges in good humor.

Amir retrieved a ruler from a workbench. "Check it with this."

Turk measured the tube. Old school. "About 1.2 inches."

"Here's your lesson for the day. Look at the part you're measuring and get an estimate of what you think the answer will be before you measure. That way, when you measure it and get a .5, you'll know that your digital calipers *weren't reset to zero*."

"Oh, the calipers," Turk said, teddy bear smile in full bloom. "I've never used them before."

"You have to trust your mind over the machine, especially when you're dealing with digital things."

Lesson delivered, Amir moved on to the shooter prototype. Daniel Huthsing and Yidi Wang had built a ramp out from its back end that led balls down just below the spinning wheel. The ramp, held together by clamps and glue, stored six balls. It worked like an upside-down Pez dispenser: When one forced a ball onto the top of the ramp, the bottom ball was pushed off the ramp and toward the newly named flywheel. The spinning wheel propelled this ball out the hood.

"I want to see how fast six balls will fly out," Amir said. Once they had a trailer in their sights, rapid-fire would be key to scoring as many points as possible.

Stopwatch in hand, Daniel counted down from three. Amir began forcing another ball onto the top of the ramp, but before a single shot launched from the flywheel, the whole prototype collapsed into a jumble of wood. Balls scattered about the room.

"That didn't work," Amir said amid his students' laughter.

Amir and the shooter crew reassembled the prototype, securing it with larger clamps. Daniel counted down again. Amir pushed the balls down the ramp. The first in line connected with the flywheel, traveled up the curved hood, and shot straight out into the trailer positioned a few feet away from the prototype. Then another. And another. Four out of six shots in less than two seconds. In the next test, the prototype went six for six in 1.2 seconds.

"How fast do you think we'll be able to get balls from the floor to the shooter?" Stan asked.

"As fast as we need," Amir said, even though he didn't have any firm idea of what kind of delivery mechanism this would be.

Stan and Amir sat down across from each other in chairs in the middle of the build room, both worried that the team had yet to settle on their overall design, let alone begun executing it. This led to a philosophical discussion on how advanced a design they should aim to achieve. As was typical of the two, they failed to see eye-to-eye.

"You and I both want to build a robot that will win," Amir said.

"No, winning's more important to you," Stan said, running his hand through the wisps of his thinning, ruffled hair. He always had a half-bemused, half-frustrated look on his face. "I'd like to have a robot that doesn't fail."

"What'd you rather have: a robot that won't have any problems and might do okay, or a robot that will kick butt?"

"I value reliability over winning."

"Okay, but if we put all of this time into this, and we get out there and look like bantamweights, it'll kill me," Amir said.

"What would kill me is if we get the robot out there and it falls over with smoke coming out of its side."

Amir nodded and then stood back from his chair. "Okay, listen up," he said to his students. "I need you all to think about how we can get the balls off the ground, into the robot, and to the shooter. You have two days."

In their kickoff-strategy meeting, the team had agreed that storing as many balls as possible was a key factor in the game. The more balls they possessed when they tracked down an opponent's trailer, the greater their scoring potential. Given this was the core of their robot and its largest mechanism, every other part of the robot would remain in limbo until they decided on its design.

Early in the first week, they had raised the idea of "the Helix" to store and deliver balls to the shooter. *FIRST* teams had used this concept in previous years. When balls were sucked into the robot, they were directed up and into the bottom of this mechanism. Think of a spiral staircase where the steps have been replaced by a ramp. Once the balls are on the ramp, two cylinders compress them. The first is a small rotating cylinder that runs through the center of the spiral ramp. The second is a large fixed cylinder surrounding the ramp. Spinning the interior cylinder would move the balls up or down the ramp in single file.

Amir was sure they could get this mechanism to work, but he didn't think it was the optimal way to store balls. One, the balls would be handled as individual units in the helix, moving up the spiral ramp in single file. Amir thought their robot would be able to store more balls if their design treated them like the molecules in a fluid. The balls would move around and be processed randomly in their system, so that the first ball collected could very well be the last shot. Two, the cylinder would have to be relatively short because the drive train and electrical board were at the base of the robot and the shooter was at the top. Three, a cylinder wouldn't hold as many balls as would a box of the same height.

As a design of last resort, Amir asked several students to start mocking up a helix prototype. They did not cobble this prototype together with spare kit parts, wood, clamps, and glue. Instead they created it on their computers using SolidWorks. This 3-D CAD software showed on-screen how the mechanism would work and how many balls it could hold.

Over the next three days, Amir and his team thought of little else than what they were now calling their robot core, the mechanism to store and deliver balls to the shooter.

Chase suggested his revolver magazine. From the intake roller, balls would feed up a conveyor belt and then drop into one of six chambers that stored several balls at once. When they wanted to shoot, a chamber would line up with the shooter and balls would be pushed upward. Too complicated, the team decided.

Amir conceived of a design where balls would feed into a platform or net that could then be raised up like an elevator to the shooter. A simple prototype his students built quickly ruled out this idea.

Several other students thought of a mechanism that resembled a gumball machine. Balls would fill a big round chamber, and there would be some kind of spinning arm that agitated them, keeping them loose. A hole would then open at the bottom of the chamber to allow one ball at a time to feed into a column connected to the shooter.

Amir liked this idea because it treated the balls in storage as a fluid. This would allow them to pack many more into a single space than if each ball was handled as a single unit, as in the helix or revolver idea.

On Wednesday, January 14, they prototyped the gumball

machine—offline. They took a sheet of regolith and rolled it into a cylinder to make the storage tank. Then they placed a bicycle wheel horizontally on the bottom to act as an agitator. When they filled the cylinder with moon rocks and spun the wheel, the balls barely budged. Their woven bands were locking into one another, jamming up. Fewer balls worked better, but they wouldn't fall through an open hole in the bottom.

The next day the students prototyped an offshoot of the gumball machine. They cut a hole in the side of the plastic cylinder. The idea was to spin the bicycle wheel, and as balls spun around in the cylinder, they would exit through the hole one at a time. As Turk, a member of the crew responsible for the robot core, spun the wheel, the balls twirled about in the cylinder but never went through the hole.

Nonetheless, Amir was excited. "Do you know why it's not working?" he asked.

They all shook their heads.

"Because it's obeying the laws of physics. The balls are moving to the outside. They're not going out the hole because of tangential trajectory. The ball doesn't want to fly straight out the hole, it wants to keep going around."

"We get it now," Turk said. "Too bad we didn't understand it last year in AP Physics."

"Ahhh, shut up," Amir said before he stomped away, exaggerating his frustration for a laugh.

That same day, Max Garber approached Amir and Stan. Max was leading the design of the computer-drafted helix. He was a member of the SolidSeven, the group of seven students on the team who had taken SolidWorks lessons taught by a local engineer before the season started. "How about a figure eight?" Max suggested. Six feet tall, Max suffered from chronic bedhead, his glasses never seemed to be quite straight on his face, and he spoke with a slight lisp. He had entered the academy his senior year as one of the replacements for the students Amir had dismissed the previous summer.

Amir and Stan turned to each other, matching lifted eyebrows and tilted heads. "What are you thinking?" Amir asked Max.

"The balls go up the middle, weaving back and forth," Max said.

"Go prove to me that it'll work." Amir thought about the concept for a moment, then beamed. "I love this idea. I want to marry this idea. But only if it works."

Encouraged, Max returned to his workstation. He wrapped a pair of wires into a figure eight, but he soon saw that its structure would take up too much room, not allowing for enough ball storage.

It was back to the helix.

Later that night Stan helped Max with his SolidWorks prototype. When they finished, Amir sat down between them, his body coiled tightly in the chair. No jokes, only questions. Several other students congregated around the computer.

"Panic mode's setting in," Stan said.

On the screen, Max illustrated fitting fifteen balls on the spiral ramp. Amir wanted to store at least twenty to twenty-four, plus he was worried about how balls would feed from the top of the ramp into the shooter. Still, Max had created a feasible design that Amir was certain they could get to work. And time for finding another was running out. "Can I have a side view of the helix?" he asked.

Max clicked the keyboard, then moved the mouse pointer over the 3-D spiral to show the different view.

"What if we move the turret to the side?" suggested Stuart Sherwin, a SolidSeven member responsible for designing the shooter. Until now they thought the turreted shooter would be positioned at the center point of the top of the robot.

"That could work," Amir said, shaking his head. Sometimes the obvious was hardest to see. By moving the turreted shooter several inches away from the center point, they could extend the helix, allowing for a few more balls, maybe eighteen total. "I'm feeling confident that if we can execute everything here, this puts us in the ninetieth percentile."

On hearing the figure, Max stared at Amir like his teacher had just failed him on a test he thought he'd aced. Max had spent a lot of time on the helix design, even though it was intended only as a backup. He had altered it scores of times, trying to fit as many balls onto the spiral ramp as possible. He thought this change Stuart had suggested would get them to 100 percent.

A student standing behind Amir asked, "Is there any way we can get better than that?"

Max sighed, but Amir was thinking the same thing.

Before the team broke for the night, Stan and Amir retreated to the machine shop to clear their heads and talk.

"We have to move forward," Stan said.

They focused on the drive train for a moment. Stan thought that a four-wheel omnidirectional drive was the way to go. He had been thinking of a design that would reduce the number of motors needed. His design would also position the components of the drive train on the left and right side of the chassis, opening up some space in the middle of the robot base.

"I like the idea," Amir said. Then he had his own insight. "Why do we need the electrical board on top of the chassis? If we move it to the side, and the middle of the robot base is now clear of the drive train, we can get a whole other layer of balls."

They looked at each other. That was it.

They left the machine shop. Amir headed straight to Max. "How many balls can you get if you start the helix on the floor? CAD it up."

Max started punching in dimension numbers in SolidWorks. On the screen, the spiral ramp expanded. By bringing the helix to the floor, Max found he was able to add another turn of the spiral. This allowed them to store four more balls (worth eight points if scored on an opponent trailer) and gave the helix the twenty-plus-ball capacity they wanted.

"Let's move on," Amir said. "We're at one hundred percent."

Max smiled. Amir gathered the team to vote on the helix and four-wheel omnidirectional drive train. It was a go. With the intake roller and turreted shooter already set, they finally had their basic design.

For Saturday, January 17, Amir had one mission for his team: to begin moving toward the final design of each mechanism in SolidWorks.

At 10 A.M., a grande café mocha in hand, Gabe arrived first. The build-room door was locked, so he waited outside for Amir to show up. His teammates began to arrive in their cars, filling up the parking lot. Through their open windows came the sounds of country music, hip-hop, rock, and classical.

"You can tell a lot about a person from hearing the music they listen to as they come to robotics," Gabe said to Luke Seale after he had pulled his old silver VW Bug into an open space near the front door. Through his window came the banter of Click and Clack, the Tappett Brothers, of NPR's *Car Talk*.

A member of the drive-train crew, Luke had wiry, curly brown hair, round glasses, and peach fuzz under his chin. A late bloomer, he was six feet two inches of skin and bones. "I was listening to techno before," Luke said, wearing his trademark beret.

"Like I said," Gabe replied, shaking his head and laughing, "you can tell a lot about a person."

When Amir arrived, he unlocked the door and pointed everyone to their workstations. "I want a quiet environment in here today. This is Solid Saturday. Most teams take weeks to design a robot. We want to get it done in three days."

Amir met with one work group after another. He gave the drive-train crew a lesson on transmissions and motor curves, firing off questions. "What's the torque?" "What's the work-per-minute rate?" He finished with "Our goal's to have the most appropriate transmission for the motor."

Next he took the electrical group outside to talk about the different motors they would use on the robot and how they would need to be wired. Then a short meeting with Max about the helix's dimensions devolved into a discussion about how Americans aren't good at math because they hate and don't understand fractions. "Which is weird because we're the ones who don't use decimals," Max said.

Amir explained, "It's self-loathing."

After lunch, the team procrastinated before returning to work. They circled around the car of Alejandro Veloz, who was also on the football team. Some cheerleaders had written on his window with lipstick, "Dear Alej, I admire you from afar, and I couldn't contain my love anymore. Your hot bod and big, smart robotics brain really get my gears going. We love you." On the passenger side, where Alej's car-pool buddy Turk sat, the window read, "Honk if I turn you on—Turk."

When everyone gathered back in the build room, Amir said, "I feel like no one understands the urgency here. I want to be like a commander on a ship. I want to shout orders and have people listen. Anyone here ever see *Deadliest Catch*?"

"Yes," a student said.

"I want it to be like that."

Amir was trying to keep things light, but he was worried about the team falling behind schedule. Instead of designing in Solid-

Works, students were taking bets on whether or not John Lennon said "Shoot me" at the beginning of the Beatles song "Come Together" and hovering over a platter of brownies that had been dropped off by one of their parents.

With the frazzled smile of a mad scientist, Amir declared, "Everyone, do you understand we're competing against teams who build robots that *build* robots?"

The Competition

THE THUNDERCHICKENS, TEAM 217

Robots. They were surrounded by robots.

Twelve members of the ThunderChickens, the 2008 *FIRST* champions, had just passed through security at the headquarters of FANUC Robotics America, a sleek modern building hugging a snow-covered hillside in a northern suburb of Detroit. Now Paul Copioli, their lead mentor and one of the company's senior engineers, was leading them across a walkway over the factory floor.

The robots resembled small cranes whose cantilevered, multi-functional yellow arms set on rotating bases could paint cars, stack pallets, assemble parts, or arc-weld steel. Their lightning-fast yet precise movements were magical to watch, and they often moved in concert with other robots, not a human in sight. There was even a glass-enclosed chamber where robots were building other robots. The dream tour inspired the students.

Soon enough Paul led them through another security door to start their work in a company conference room. Everyone took a seat as Paul stood by the whiteboard. He was an imposing presence. He had the big, rounded shoulders of a former college swimmer. He always looked one straight in the eyes and had the habit of leaning forward on the balls of his feet like he was preparing to pounce. The hard-charging, take-no-prisoners attitude of the legendary ThunderChickens was all Paul, an Air Force Academy graduate. He was the kind of leader one would want to go into battle with: whip-smart, focused, forceful, decisive, always thinking three steps ahead, and brutally honest. In the *FIRST* community, people called him both a robot god and a human Roman candle.

It was five days after kickoff, and Paul wanted his design team to have complete, manufacture-ready blueprints for their robot, from top to bottom, in two weeks. They did not balk or doubt their ability to accomplish the goal. That was not the ThunderChicken way.

With more than 1,600 teams in the *FIRST* Robotics Competition, diversity was not a problem. Teams represented almost every state in America and every demographic. Small rural towns and big metropolises. Wealthy suburbs and the worst inner-city neighborhoods. Ten other countries also fielded teams, including Israel, Canada, England, Brazil, Mexico, and the Philippines. Some teams had a handful of students with a single mentor, often a high school teacher, to lead them. Others had more than seventy students and a dozen mentors, all with professional engineering experience. Rookies often just wanted to field a robot at a regional competition. Many others were content to do well enough in qualifications to reach the tournament elimination rounds. Then there were the elite teams that expected to win a regional competition every year, earning a spot in Atlanta, where they were almost always in the running to claim the championship. These select teams were well funded and super dedicated, and competed almost year-round in robotics competitions.

The ThunderChickens were one of these teams. Ford Motor

Company sponsored them, and year in and year out, they built a finely engineered, dominant machine. They had lots of students, lots of mentors, and lots of supporters. Excellence was expected. Victory was the standard.

They were not shy about their motives. "Success is an addiction," Paul would answer to anyone who cared to ask about his drive to win. He aimed to get his kids hooked on success. The more the ThunderChickens won, the more the students contributed, the more they drew from their time on the team. The success they experienced on the ThunderChickens might just keep these kids interested in engineering, Paul thought. That success might one day help save Michigan. That success might one day help save the country.

At the start of the 2009 *FIRST* season, the housing crisis continued to sweep the country, Wall Street had imploded, and the United States economy was in meltdown. General Motors, Chrysler, and Ford teetered on the edge of collapse—and with them Detroit. President George W. Bush had thrown the first two a lifeline of $17.4 billion, but all three were burning through billions a month, and they did not know what president-elect Barack Obama would do once he took office on January 20. The uncertainty over the future was bad, but worse still was the present-day reality of sinking car sales, spiking operating losses, plant and dealership closings, reduced hours, work stoppages, early-retirement buyouts, benefits cuts, and lost jobs. Every week something new hit, and it seemed like a rain of blows in the advance of an imminent knockout. The automobile industry, once looked on as the economic giant that created the middle class, now symbolized everything that had gone wrong in America.

HELP a red spray of graffiti pleaded on the side of a building by a shuttered assembly plant in Sterling Heights, home of the ThunderChickens. IT DOESN'T EXIST came the answer below in a shock of green paint. The auto industry was the lifeblood of this community of 128,000, located on a square, flat, featureless stretch of land in Macomb County, 20 miles northeast of downtown Detroit. Hope ran in diminishing supply, and bitterness seeped in to replace it. A popular bumper sticker read: OUT OF A JOB YET? KEEP BUYING FOREIGN. Chrysler and Ford had two remaining big plants each in Sterling Heights, and one could not drive for more than a few minutes in any direction along the city's four-lane roads without running across an auto-related supply plant or office complex. The vast empty parking lots signaled that the halcyon days had long passed. The surrounding chain restaurants and big-box stores were never crowded. Even the gas stations in a city that loved its cars rarely had a line. In the subdivisions of modest homes, FOR SALE BY BANK signs often stood in rows on front lawns, this in a community rated one of the best places to raise a family only the year before. In the midst of this crisis, Richard Notte, the mayor of Sterling Heights for sixteen years, admitted, "There isn't any light at the end of the tunnel right now."

Every member of the ThunderChickens had been touched by the crisis. Paul was about to lay off several of his staff. The handful of present and former Ford engineers who served as mentors were worried about their retirement. The parents of the forty-two students on the team feared losing their jobs or had already been downsized. Some were taking second jobs, flipping pizzas and the like. All were tightening budgets and struggling over how to pay for college.

However, there was light at the end of the tunnel for some in Sterling Heights. Paul felt that those on the ThunderChickens were well prepared for the future. "It's the kids that *aren't* here that I'm worried about," he said. One student's father, who had gone to work on an assembly line straight out of high school and had run out of choices, echoed this sentiment. "I want my son to be better than I was. He will be," the father said, explaining that his son's three years on the ThunderChickens would give him a leg up on the future.

The ThunderChickens met their deadline. Three weeks into the build season, they had completed manufacture-ready blueprints for their entire robot. Their design mirrored that of the Dos Pueblos team, but instead of an omnidirectional drive train, they were going with a version of the six-wheel tank drive they had used with success in previous seasons. They had also decided not to have a camera-tracking turret, relying on their robot pilots to zero in on opposing trailers.

Each part—from the drive train, to the intake roller, to the spiral to deliver balls, to the turreted shooter—had also been prototyped and refined. This work was done at their base of operations, the Ford transmission plant on Van Dyke Avenue and 18 Mile Road. They even had all these prototype mechanisms working together. Albeit built out of plastic, wood, and a selection of parts from the kit, this prototype robot surpassed what many teams would bring to the competition as their final model.

They worked a lot of hours, but this was not the reason they were so much farther along than most *FIRST* teams, who would struggle up to the last hour to finish their robot by the February 17 deadline. Their progress was the result of the Thunder-Chicken version of military efficiency. Paul was the general who had the knowledge base and experience to know how and where to lead his team. The Ford engineers ran what amounted to a boot camp on how to assemble a transmission, build a drive train, wire the robot, and fabricate the parts. A clear hierarchy of command existed among the students, with sophomores known as grunts and designated seniors acting like officers, their word as good as a command. They had tactical plans for each day, with students separated out into different details, their duties known. They even had their own uniform, neon-lime-green T-shirts that could be seen in the pitch-black of night. Foremost, they had a clear strategy: Overwhelm. Dominate. Win.

The designs completed at FANUC would soon be sent to one of their sponsors in Texas, who would fabricate two identical sets of parts in their industrial machine shop. One would be for the ThunderChickens' practice robot, the other for their competition robot. Once these parts arrived at the Van Dyke plant, the team would unpack and organize them. Then the students would form an assembly line in one of the long, empty cavernous rooms that had once housed Ford staff during the boom years. Massive blueprints for each mechanism would be laid out on the tables, the relevant parts set by their side.

Then the students would put together their robots. They would still need to machine various parts at the plant, but the bulk came from their sponsor. By week five they would have a fully operable practice robot. Several days before ship, they would have a finished competition robot.

With so many students and mentors, all of whom were very good at what they were doing, there was enough time for the team to help assemble robots for the two teams the ThunderChickens were mentoring that season. This was part of the *FIRST* ethos: experienced teams guiding rookies through their first build.

After the ship date, their drivers would spend many hours running drills with their practice robot and scrimmaging against other teams on a regulation-built field in the local area. Paul often repeated that there were three pillars of success in *FIRST*: strategy, the robot, and drive practice. By the time they reached their first regional in Chicago, the three would be more than set, and the ThunderChickens would be prepared for victory.

Week Three

29 DAYS UNTIL ROBOT SHIP

I'm writing code. Ignore the fact that my back is to the computer.

—GABE RIVES-CORBETT

On Sunday morning, January 18, Luke Seale entered the Bat Cave. That was the nickname of Stan Reifel's office in his Santa Barbara home, where the SolidSeven did their intensive design work—it sounded cooler than saying, "I'm going to Stan's." The dimly lit space with its drawn blinds allowed for day to pass into night without noticing. Radio-controlled foam airplanes hung from the ceiling. Electrical gear, including soldering equipment and crimping tools, filled one side of the room. A large desk stacked with schematics of circuit boards occupied the other. On an adjoining table stood an old microscope and an antique Molle typewriter. Computer-programming guides and mail-order-parts catalogs lined the bookshelves. Like Stan, everything here was tidy. Even though he was married and had a young daughter, one had the feeling that he could spend weeks buried in a project in this office and never come out. The Murphy bed along one wall suggested he had.

In 1983, Stan had graduated from UCSB with a degree in electrical engineering and taken a job developing robots at a research institute in Menlo Park, California, near Stanford, the epicenter of all things future. After five years, he started his own business writing software for a biotech firm. From there, he spent a stint in Boston building robots for a pharmaceutical company, then in 1998 launched an Internet company selling custom-printed circuit boards to electronic enthusiasts.

Now fifty, Stan had made enough of a fortune to retire whenever he wanted, but besides his black convertible Mercedes, he showed no signs of his wealth. His house was nice but by no means extravagant. He almost always wore an off-the-rack button-down shirt, faded jeans, and sneakers. He carried himself with a kind of Howdy Doody, easygoing manner, except in those rare moments when his smile tightened at the corner of his mouth and his face reddened.

The previous year had been his first with Team 1717. He had heard of Amir through some friends and volunteered to mentor the students in part because his workday had become dominated by business, at the expense of engineering. Stan wanted an outlet to tinker and build, and he wanted to help nurture the next generation of engineers. He did not like pressure, nor was he remotely competitive, which made for an odd pairing with Amir.

The relationship had experienced growing pains their first year. Amir had previously brought in mentors to help with the electrical and programming aspects of the robot because they were not his areas of expertise, but he had never had outside help with what he considered his domain: mechanical engineering.

Throughout the first few weeks of the 2008 build season, Amir kept a close watch on Stan, making sure his work with the students was up to Amir's standard. Midway through the build, Stan had reached his breaking point. "I don't feel respected, and you micromanage everything I do. You are a great teacher and a good engineer, but you just don't know how to manage people. You want to control everything, and I don't feel like you trust what I'm doing."

Amir was taken aback. "I need to be able to work with other people who can help me run this program as it grows. If I can't make this work out with you, then there's not a lot of hope."

"I appreciate your sentiments," Stan said. "I know I'm stuck in my ways, and I'm not going to change. I don't think you'll be able to change either."

Amir looked at the mentor incredulously. "Stan, I'm not a very religious person, and I pretty much believe that we've one shot to get this right. If you're telling me that I can't change, that I can't evolve as a person, then that's a pretty bleak outlook. We've all got a limited time on this earth to get it right, and I believe I'm capable of changing."

"Okay, we'll see," Stan said.

For the rest of the 2008 season, Amir forced himself to loosen the reins on Stan and trust him more. There were moments when he caught himself slipping back into micromanage mode, but the season had gone well enough that Stan came back for his second year.

Before the 2009 season, they had agreed that Stan would work with individual students on designing some of the more involved mechanisms on the robot. The drive train was top of the list, and Luke was in the Bat Cave to model it out in SolidWorks. Of anybody on the drive-train crew, including Chase, Luke had the most experience with the design software. He and Stan sat side by side in front of a large flat-screen monitor in the corner of the office, working from a sketch that Amir had presented to the drive-train crew the day before.

Amir and Stan had roughed out the design themselves, its sophistication far beyond what the high school students could do on their own. But this was what distinguished the *FIRST* competition from a science fair where students took what they already knew and put together a project on their own. By intention, Dean Kamen had no rules about the level of mentor involvement, allowing the kids to get real experience working hand in hand with engineers.

Luke was now about to learn how to take a conceptual sketch, break it down to its individual working parts, design them to precise measurements for machining, and see how they meshed to form a functioning, efficiently designed mechanism. As Chase said, it was the difference between figuring out how to put together a go-cart on one's own versus learning how to build a Maserati with help.

The drive-train design consisted of four identical wheel modules. The most basic element of the module was the slick kit wheel that all teams had to employ on their robots. This was where Stan and Luke began in SolidWorks.

"Draw the wheel," Stan said.

Luke started clicking through menus on the screen, his tall, narrow frame going slack in his seat. In a few steps, SolidWorks would allow him to design a wheel in three dimensions that matched the ones provided by *FIRST*. He would start by drawing a circle on the screen and punch in the diameter of the kit wheels. Then he would stretch the circle into a cylinder, then bevel the ends of the cylinder to create the rounded edges of a wheel.

As Luke and the other SolidSeven had learned, SolidWorks is powerful CAD software that makes it easy to create and manipulate geometrically complicated parts on the screen before fabricating them. Professional engineers use the software to design everything from the Mars rover, high-performance race cars, and planes to stereo systems, toys, and Trek bikes. With its power came a complex and innumerable variety of tools and functions. Realizing the potential of the software often became frustrating, particularly when trying to mesh parts together, like the wheel and axle. Then SolidWorks could be merciless. Sometimes the program seemed like it was waiting for a smarter, more capable person to come along who wasn't trying to fit a square peg into a round hole. Since Stan was practiced at SolidWorks but by no means an expert, he and Luke sometimes had to stumble their way through the software together.

After the wheel and axle, they drew a large pulley and attached it to the wheel. Then they added side plates for the module that held these parts together. The large pulley needed to be powered, so Luke drew another horizontal axle mounted to the side plates above the wheel. This axle had a smaller pulley that would be connected with a timing belt to the large pulley.

They had designed the sizes of the pulleys so that every time the small pulley made three rotations, the large pulley would make a single rotation. This created a gear ratio of 3:1. Like the chain and sprockets on a bicycle, the belt and different-sized pulleys were a transmission, which traded speed for torque. When the whole drive train was complete, with two pairs of modules connected together (each with another transmission) on either side of the robot, they determined a total gear ratio of 6:1. They thought this would provide the optimum speed/torque relationship to move about the low-traction regolith field.

Since the module also needed to be able to pivot in any direction, the design would require the complexity of axles rotating within axles, ball-bearing blocks on ball-bearing blocks, and more pulleys and timing belts.

Stan would tell Luke the approximate size of a part they needed within the module, then go off to make sandwiches or get Cokes to keep Luke moving forward. Stan would typically return when the part was almost finished. He would show Luke where screw holes needed to be placed or question how this new part would fit with the others. Sometimes Stan would start his suggestions, like the one where he wanted Luke to punch out sections of the aluminum pulley to cut down on weight, by saying, "Now, Mr. Shaeer is going to want..." He felt better asking for the additional work by placing the blame on Amir. Both Amir and Stan played this game, and the students were wise to it but never said a word.

Stan and Luke frequently searched online for data and sizes of various pulleys, ball bearings, gears, and timing belts. Based on what was available to be ordered, they changed their own design. The process was iterative, and they continued to revise and add to the wheel-module design until everything looked just right.

When they completed the wheel module, Luke could not feel his wrist after ten hours clicking away at the mouse, but he was upbeat. On-screen, their three-dimensional design looked elegant and refined. The module consisted of forty parts that they had designed and assembled, all in SolidWorks. With the exception of the wheels and ball bearings, every one of these parts would need to be fabricated or modified from off-the-shelf parts in their machine shop.

"That was a good crash," Gabe said, sitting in front of his laptop. He stared at the lines of software code running down his monitor, somehow able to see what he needed in the blur of seemingly random letters and numbers. It was inauguration day, January 20, but besides a few kids wearing Obama T-shirts, it was all robotics, all the time.

"A good crash?" his fellow programmer Kevin asked while standing in front of a camera mounted to a motorized rotating tripod. In his hand he held a board stapled with large square swatches of bright pink-and-green fabric. This board was meant to simulate the pink-and-green vision target at the top of the center post in each trailer, and they were trying to program the camera to track it.

"But the camera's tracking two million targets," their third compatriot, Nick, said, trying to understand why Gabe was not more defeated by the error messages filling his computer screen.

Gabe explained that the tracking code he had written had worked up until the point where there was no code at all for the computer to follow. Only then had it crashed.

Nick and Kevin took Gabe at his word. Of the three, he was by far the more accomplished and experienced programmer.

Since the season began, the programming crew had sat on the edge of their seats, eyes glued to their laptops. They bickered constantly, and every stage of their attempt to get the camera to track the target was either a great success or a tragic failure, with no in-between.

Later that afternoon they ran to get Amir to show him one of their successes. Amir looked at the screen where a red outline of a box shifted up and down, following the movements of the cloth-covered board in front of the camera. He wasn't sure what kind of breakthrough they had made, until Gabe announced: "It's tracking colors like it's supposed to, and nothing is exploding."

A few minutes later Gabe ran into more problems with his code. He stood away from his chair, kneaded his eyes with the palms of his hands, and said, "I'm going home."

Soon enough he returned to his workstation. He knew well that without their code, the robot his teammates were building would sit lifeless on the field.

FIRST provided each team with a robot controller designed by National Instruments and used by engineers and research scientists around the globe. Placed on the robot to serve as its brain, this controller was basically a rugged, high-performance computer, but it needed software. Otherwise, the robot was like a newborn baby, without any knowledge of how to navigate the world around him. The programmer was responsible for supplying the robot with this knowledge through sets of precise instructions.

There were two basic approaches to controlling the operation of the robot: open loops and closed loops.

To understand an open loop, think of a shower. You rotate the dial to a nice, pleasant, warm cascade of water. Everything is wonderful until someone flushes a toilet in the house. Your skin then gets scalded unless you're quick enough to turn the dial yourself. No sensor measures water temperature, nor is there a mechanism to maintain it when disturbances, like a diversion of cold water to the toilet, occur. The loop is open because it depends on human input to make changes or compensate for any circumstances that affect the system's output.

A closed loop doesn't require human intervention. A modern household thermostat operates on this principle. On the first chilly day of the year, the house feels cold, and you set the thermostat to 70 degrees Fahrenheit. The closed loop begins. The heater fires up. The thermostat constantly measures the temperature in the room. When the heater brings the temperature to 70 degrees, the thermostat shuts down the heater. If the temperature falls below 70 again, the heater turns on again. The thermostat relies on a sensor, the temperature gauge, to provide information—or, more precisely, feedback—on whether or not it needs to fire up the heater. Sensors are the key to the closed loop.

At their most fundamental level, these loops are based on statements of logic that are either proved true or false. Imagine trying to get a very basic robot to navigate around a room's perimeter on its own, turning each time it encountered a wall. To start the robot, the controller runs an open-loop logic statement that essentially says: *If switch is put into "on" position, drive forward;*

otherwise remain in place. Given only this instruction, however, once the switch is on, the robot will drive forward, hit a wall, and keep ramming against it, ad infinitum.

This is where the closed loop comes into play. Say the robot has a short arm out front with a sensor button that depresses when it comes into contact with a solid surface. Once the robot is driving forward, the controller runs a closed-loop statement that says: *If the button is depressed, stop, turn right 90 degrees, and drive forward; otherwise drive forward.* With these instructions, the robot will turn once it encounters a wall and drive along the room's perimeter.

These statements, running continuously within open and closed loops, were the building blocks on which the entire robot would operate, from the drive train, to the ball collector, to the helix, to the turreted shooter. There would be separate sections of code that controlled each of these mechanisms. The three programmers called these sections "managers" (as in drive-train manager). Each needed hundreds of logic statements to deal with every situation and circumstance.

Even the simplest mechanism involved a lot of programming— for instance, the code to spin the shooter flywheel. Gabe had to write a loop for the robot controller that ran thirty times every second to verify if there were any inputs coming in wirelessly from the joystick. When the controller detected that a robot pilot had pressed the joystick button to rotate the shooter flywheel, it sent this information to its shooter manager. On processing this information, the code in this manager resolved to provide power to the flywheel motor.

The shooter manager then sent a signal to a device called a speed controller on the electrical board. The speed controller operates like a faucet, controlling the flow of electricity from the battery. On receiving instructions to power the shooter flywheel, the speed controller "opened its faucet" just the right amount, allowing a flow of electricity to the motor that spun the flywheel. Until the shooter manager informed the speed controller differently, the current would continue to flow.

Programming the controls of other mechanisms, like the rotation of the turret, entailed a whole different realm of complication. For instance, the rotation of the turret required a closed-loop system that needed sensors to detect its location. Instructions to move the turret from where it was to where it should be necessitated high-level calculus concepts that added or subtracted power to its motor until the angle of the turret matched the angle of the joystick.

It was important to remember that Gabe was seventeen, and that before the season, he had never programmed a robot before.

But Gabe had a lifelong affinity for computers—literally. His mother delivered him at home, and he was already pawing at a desktop computer hours after he came into the world.

His father, Lionel Corbett, was born in England, the son of a German Jew who escaped the Nazis and changed his name from Gorbitz once he landed on British soil. Raised in Manchester, Lionel was pushed into medicine by his parents. In 1972 he moved to the United States, specialized in geriatric psychiatry, taught at various universities, and wrote long, involved books on the interplay between psychotherapy and religion. He was sober, contemplative, and probing.

Gabe's mom, Cathy Rives, graduated high school a year early in Atlanta, Georgia. She married at seventeen and started a theater company. After a quick divorce, she traveled the country in a VW van, married again, ended up in California, and divorced again. At twenty-eight she enrolled in medical school at Rush University in Chicago, where she met Lionel. He was forty-three at the time, twice divorced, and head of the psychiatry department where Cathy was a resident. They married, prompting a university scandal. Gabe, their only child together, was born in Sante Fe in 1991. Soon after, they moved to Goleta, where Lionel and Cathy divided their time between practicing psychiatry and raising their son. Inevitably, the two intermixed.

Cathy and Lionel believed that a child, whether three or seventeen years old, deserved to be respected and treated like an adult. As a toddler, Gabe did not have a fixed bedtime. As a teenager, he would write down his dreams on a notepad by his bed, and his dad would analyze them. When Gabe left the house at night, his parents asked when he would be home rather than giving him a specific hour.

After Lionel was informed he needed a bone-marrow transplant, he and Cathy sat Gabe down, spoke about his prognosis and the possibility that he could die while in Los Angeles that winter.

This kind of treatment gave Gabe a maturity well in advance of his years but made him averse to environments where he was not given the respect his parents showed him. He experienced this most in his education, where he transferred between a few elementary schools until he found one that was less authoritarian and gave students responsibility for their own learning.

The other issue in his education was that Gabe often found himself bored. At four, he was reading proficiently and imagining that the pots, pans, furniture, and piles of pillows he had linked together with string were "electricity machines." At eight, he was well past what they were teaching him in math at school, so his father taught him algebra himself. In junior high Gabe never needed to study, able to read a text once and ace the test.

Computers offered a world where he was never bored. He was already writing simple programs on his father's computer when he was just five. When he turned eight, his parents bought him his first computer. They had rarely seen such joy. A few months later, his parents hired a tutor from UCSB to teach him programming. Gabe closed himself in his room and spent whole days and weeks at the keyboard, writing increasingly involved programs. He loved the math, logic, and creativity of it. He loved those moments when he felt one with the computer, able to communicate and process information as it did. By fourteen, he was collaborating online with software engineers.

When the robotics season began, Gabe figured he had already spent twelve thousand hours writing code. This was two thousand hours beyond the amount of time that, according to Malcolm Gladwell in his bestselling book *Outliers,* Bill Gates had needed to become an exceptionally talented programmer.

But Gabe's talents didn't stop at the keyboard. He operated with the attitude that he could excel at anything. He went sailing one day, found he liked it, and became a certified sailor. During a cruise one summer with his parents, he took on scuba diving and became a certified diver. He wanted to help the disaster relief during the many forest fires in the area, so he became Red Cross certified to work in shelters. His next goal was to become a certified paraglider. Each certification was like a badge of honor.

Gabe also excelled at tae kwon do. He didn't much care for the fighting, but he liked the community he found in the dojo and mastering the techniques. He rose to a half-red/half-black belt, one level below the highest stage, and had already competed three times in the Junior Olympics.

Finally, there was theater. In seventh grade he wanted to act but he didn't win any roles, so he volunteered to work backstage. He found he liked doing the lighting, building the stage, and running the soundboard. Theater production was art and technology combined, and he was hooked.

Sometimes Gabe wore his smarts and accomplishments on his

sleeve, but this seemed more a defense mechanism than pride. When he first arrived at Dos Pueblos, he was socially awkward and insecure. This was a kid, after all, who had programmed in his room on average three hours a day, 365 days a year, since he was five years old.

His freshman year at high school was, as he put it, "bad." He had no friends, so he would program for hours after school. Online multiplayer video games were his social network. Kids teased him because of his looks, particularly the large birthmark on his neck. It was a heap of hurt, and Gabe, consciously or not, combated this by making himself superior and indispensable, whether at his dojo, the theater, or the engineering academy. By the start of robotics season his senior year, he had overcome these insecurities. Now he was just good.

During eighth grade he applied to the academy because of his interest in math and science. At the interview, Amir laid out a bunch of material on the table: a black plastic rod, some spring wire, a brass ring and block, and some foggy, textured plastic. Alongside these materials he placed a fancy magnifying glass (the same one Amir had made for Emily when they were dating). He then asked, "What do these have in common?" Gabe had no clue, but Amir hadn't expected him to know that the magnifying glass was built from these raw materials. He only wanted to see if there was wonder in a student's eyes on the reveal. "Oh, yeah," Gabe said, the wonder there.

In his first class with Amir freshman year, Gabe found a teacher unlike any other he had known. Most teachers had an attitude, Gabe said, that boiled down to the following: "You're the student. I'm the teacher. I'm smarter than you. Now listen to me." In contrast, Amir never spoke down to him, and he always made it clear why they were learning what they were learning, even if it was "because it's on the AP test and *that*'s the reason." Academy projects, like the one where he built a speaker out of coils of wire, paper plates, and magnets, were fun and interactive, and Gabe saw how they helped him understand physics. Now with the capstone course, he was working with real engineers and doing real programming.

He figured this was the first and last time he would do so. He had already been accepted early admission to New York University's theater program and was sure theater production was his future. Maybe he would integrate some of his programming skills into designing lighting systems, but that was about it.

Still, Gabe was driven to build a superior robot, both for his own sake and so that he could help Amir succeed in his vision. "He has lofty goals, changing the face of education for the entire country," Gabe said. "How do you *not* respect that?"

Amir and the team were depending on Gabe, and he felt the pressure acutely. Every night since the build season had begun, he seemed to come home later and later, then stayed up in bed thinking about how to solve one problem or another with code. He kept a notebook by his bed in case he had a breakthrough mid-slumber and needed to jot down some code. The long hours were already beginning to wear him down, and the previous weekend he had missed visiting his father at the hospital in Los Angeles for the first time.

Amir had set a standard for Gabe that surpassed his already high expectations for himself, and he would be tested to meet it.

"What are you doing, Gabe?" Chase asked.

Gabe was sitting on the worktable, feet on a chair, his computer behind him, staring off into space. "I'm writing code. Ignore the fact that my back is to the computer."

Late on inauguration day, Gabe was finally making strides

in programming the turret-mounted camera to track and aim at a trailer automatically. Without this ability, Chase or his co-pilot, Kevin Wojcik, would have to eyeball where to position the shooter. If their robot was across the field in a mêlée of other robots and trailers, their aim would be guesswork at best. If they automated the turret, however, all Chase or Kevin would have to do was point it in the general direction of the opposing trailer, then press a button to lock on the target, much like a fighter pilot locking a missile on a moving target. The turret would then make slight adjustments until it was aimed for the shot. Even if the opposing trailer continued to move, the turret would automatically follow it, ready for the drivers to fire.

For Gabe and his fellow programmers, their initial obstacle had been writing code to process images from the digital camera fast enough that their target was not gone before they had a chance to score. *FIRST* provided out-of-the-box code to identify a target with the camera, and for the past two weeks, they had attempted to manipulate this code to speed it up. Several times they improved the code and asked one another, "Do you think we can live with this speed?" The answer was always "No."

Other factors had been throwing off the camera as well. For one, the lights in the build room were washing out the images, just like the sun did to photos taken outside. This made them unreadable by the computer. Gabe and Nick had approached Amir the previous week to ask if the lights at the competition would do the same.

"It's going to be like walking in the sun on the field," Amir said. "What you should do is find out what kind of lights are used and see—"

Gabe interrupted, "They're PARNels, and they're halogen."

Nick laughed at how insane it was that Gabe would know this

off the top of his head. Amir just looked at him and shook his head.

"I do a lot of theater lighting," Gabe said.

With these problems continuing to plague them, Gabe had decided to scrap the out-of-the-box code and write his own. It would be more streamlined, and he'd be able to account for interferences better. This began on inauguration day and continued in a rush over the next two days as well.

His code had the camera take dozens of digital images every second of a trailer's vision target. Depending on one's team, the trailer hitched to the robot had either a pink band above a green band on its center post, or green above pink. By analyzing a single pixel from the center of each band to see which color was on top, the code determined whether the camera had locked onto an opposing trailer or one from its own alliance. This ability would, Gabe hoped, keep the team from shooting inadvertently on their own alliance trailers.

Then came the difficult work—aiming the shot. Gabe thought of the turret-mounted camera like it was a person. In basketball, if you want to throw an overhead pass to a teammate who is off to your right, you turn your head and torso until your teammate is in your line of sight, then make the toss. It was the same with the turret. He wanted to rotate it until the target on the trailer was centered in the camera's field of view.

In basketball, you also need to judge the distance between you and your teammate to adjust for the strength and angle of the throw. How could the camera code judge distance? Size was the key. The farther away the target was from the camera, the smaller it would appear.

Assisted by Kevin and Nick, Gabe ran a bunch of experimental tests, taking pictures of the target at different distances from the camera and seeing how big it appeared in the overall image. Using this data they were able to write a program to accurately determine the distance of the target.

By the end of the night on January 22, three days after beginning their own code for the camera, the programming team was able to acquire images, process them quickly, and send the analysis of the target's position to the controller's turret manager.

For the rest of the week, they worked on precisely rotating the turret to the position it needed to be in to make the shot. They still needed to resolve how fast to spin the flywheel and at what angle the retractable hood needed to be set in order to shoot at the correct distance. The thought of everything that was left to do pained them.

"Our code needs to be smarter," Gabe said after the camera on the motorized tripod spun out of control for no reason.

Nick punched some numbers into his calculator. "We just need more data."

"I don't mind problems that I introduce," Gabe said. "I mind problems that introduce themselves."

While the programmers spent the whole week focused on the shooter, the rest of the team was scattered about on different projects.

In the machine shop, students were starting to produce parts for the drive train. Amir advised two members of that crew, Saher Hamdani and Lisa Nakashima, on how to make the small, cylindrical aluminum spacers that would be used when attaching the pulley to the wheel. He showed them the difference between the lathe and mill. On the lathe, the part being machined spins while a cutting tool is forced up against it to remove material. On the mill, the cutting tool spins and the part is brought up against it. He explained that they would be using the lathe because it was better for creating cylindrical parts.

Amir then showed them how to make their first cuts on the lathe. "If you do something that you don't think is right, the worst thing you can do is spastically overcorrect. If you jerk this around, you could make it worse. If something bad happens to me, I go into slow motion. When I was a kid, my younger brother was in the kitchen making some food. There was a carton of eggs on the stove. All I heard was him frantically calling my name. He was four, so I was about nine. I walked in, saw the flaming carton of eggs, calmly grabbed it and put it in the sink, ran the water till the flames were out, then went back to watching TV as if nothing had happened. My brother still remembers that to this day. The life lesson: Panicking makes things worse. Stay calm."

Once Amir left, they donned their safety glasses and began to hollow out the center of the spacer. With each turn of the handle, little spirals of aluminum fell to the floor. Throughout, Saher repeated, "Okay, no eggs on fire." A day later, the self-named "Ladies of the Lathe" came out with a cosmetics bag containing more than fifty spacers that the team would need for the wheel modules.

In the cafeteria, Amir watched Turk practice his shooting. Chase drove the 2008 robot with a trailer hitched to its rear, while Turk attempted to drop as many shots as possible into it within ninety seconds. The low ceiling made it difficult to put the necessary arc on his shots. There was a lot of trash talk, particularly when Bryan Heller, Turk's main competitor for the shooter position, alternated rounds. Turk was making almost 80 percent of his shots until Chase began to mix things up by starting and stopping the robot. Then everything fell apart for Turk. His missed shots ricocheted off the trailer and rolled under lunch tables.

"Keep practicing," Amir said.

In the build room, he then met with the electrical team. Much of their work would not come into play until the robot was ready to be wired. Right now they were putting together the operator consoles for their pilots, Chase and Kevin. Each console had two joysticks secured to a plastic box that the students had made. When Chase came over to check out their progress, he took the joystick, rotated it around, and said, "This is going to be awesome." Gabe added that he was looking into a company that had F-16 combat flight-stick replicas.

"What about the ones that just cost fifteen bucks?" Amir asked.

Gabe and Chase just shook their heads.

Then Amir was off to check on the SolidSeven, who were designing the helix, chassis, and shooter. He ran through questions about how the geometry worked, where everything would fit, how the balls would move between the mechanisms, and how they would translate the design into cutting sheet metal. They were getting closer on all these designs, but they were by no means set.

"I'm sorry it's been so frustrating," Amir said to Stan, who was overseeing all these SolidWorks designs.

"Ninety-nine percent of it's easy," Stan said. "But that last percent makes you think you don't know what you're doing."

"After we get the chassis done for the practice robot, I want to get the wheels on, add the electrical board, and start driving. We can always make changes to the competition robot."

Stan cast a long look at Amir.

"*Please* don't tell me you think we're only making one robot," Amir said, his every word almost a sigh.

"No, I'm all for making two," Stan said. "But if we don't step it up a little bit, I'll start thinking that."

Amir stopped by Turk, who looked now to be diligently at work helping Max with the helix design to deliver balls from the intake roller to the shooter. On closer inspection, Amir found that in fact he was on a website designing a basketball jersey. He had replaced an NBA logo with the Team 1717 logo on the front and was testing out different fonts for TURK to be placed on the back. The jersey, he explained, would look good on him at the competition.

"We've officially lost our minds," Amir said.

The team was pulled in other directions as well. Thursday night they presented the academy to a herd of eighth-graders and their parents in the Dos Pueblos gym. They were preaching to the converted. The eighth graders, both girls and boys, looked starry-eyed as Gabe drove around the 2008 robot and raised its forklift toward the ceiling. As for the parents, they instructed their kids on how best to present themselves to Amir, hoping it might make a difference during the admissions process. "Shake his hand. Look him in the eyes," one father urged his son.

By the end of the third week, everyone on the team was tired. Saturday night, two hours past their supposed 9 P.M. quit time, Amir lay down under one of the workbenches, his arm over his eyes, needing to escape the glare of the fluorescent lights. Chase was pale, his eyelids heavy. Gabe kept shaking his head to keep himself awake. Along with a handful of others, he and Chase were staying later and later, while the rest of the team left at nine. Resentments were building between these two groups.

"We want to start getting closer to having the real robot look like a real robot," Amir said encouragingly at the end of the Saturday-night session, his voice hoarse.

He was pleased at how much focused effort this class was putting into building the robot. Some kids were stepping up, and the divisions that had plagued the team at the beginning of the season had eased. He recognized there was a split now between those putting in the long hours and those not, but this happened every year. Over the next few days, he would remind everyone that they would get out of the robotics season what they put into it. If the students wanted to see their competition robot ready for ship, everyone needed to be working together at the same high level of intensity.

Even then, given that they still needed to design the shooter and had yet to assemble even two parts of their robot, they would be in a rush to finish.

Week Four

22 DAYS UNTIL ROBOT SHIP

First, I'm going to teach people how to fish, then I'm going
to come fishing with you, okay?

—AMIR ABO-SHAEER

*T*he phone was ringing.

On Monday, January 26, Amir was in the middle of a physics lecture. Whoever was calling didn't get the hint that he didn't
want to pick up the classroom phone. It kept ringing. Finally, Amir
stopped, went to the wall behind the whiteboard, and picked up.
He refused to get a cell phone, feeling like he was already on
call constantly, so this was the only way people could reach him
during the day. "Hello," he said, frustrated at the interruption. His
students began chattering away, their focus lost.

It was Parm Williams, one of the oversight-committee members for the Women's Fund of Santa Barbara. The organization
was a potential donor. "We're giving you the grant," she said.

When the words sank in, Amir wanted to jump up and down,
but he had to play it cool in front of his class. The grant was critical, not only for the funds but also because the Women's Fund
represented validation from a major donor that could open up
more doors for fund-raising. Up to this point, most of their donations had been from parents of his students. "Thank you," Amir

said formally. "I'm really appreciative of your time and the effort you put in presenting the academy to your group."

Parm told him the academy would receive $150,000, an amount double that of any of the Women's Fund's other donations that year. They liked the academy, particularly how it was engaging young women.

Amir returned to his class, eager to tell Emily the good news as soon as he could. They were still a long way off from matching the $3 million state grant by the November deadline, but this was an essential leap forward.

Later that day, he met up with his team in the build room. He couldn't tell them about the grant, since it was not yet official, but he had another secret to share with them. "Okay, listen up," he boomed, coming out of the machine shop. His strong voice masked his discomfort. His body language did not. He held his hands up, palms out, as if to stop himself from what he was about to say. He glued his eyes on the carpet several feet ahead of him. "I've got a personal announcement to make."

Amir often shared stories of his own life as a way to teach, and this also allowed his students to feel like they could talk to him about anything. They were always curious to know more. This year, his team had unearthed old Dos Pueblos yearbooks, giggling at pictures of Amir in his high school band uniform. They also liked to interrogate him about his lifestyle, like why he was a vegetarian.

"It's the way I was born," Amir had explained to the South Korean–born John Kim, who couldn't quite understand why his teacher would refuse to eat meat. "It's the way I came out of the womb."

"So," John said, "you came out saying, 'I want *leaves*'?"

But there remained parts of Amir's life that he wanted to keep for himself. Their baby, still early in its second term, had been one of those, but with Emily about to show, they felt like he had to share now. Everyone was rapt with attention.

"I'm going to be a father," Amir said.

Claps, cheers, and several shouts of "I knew it!" filled the room. If one thing held the team together, it was Amir.

"I just wanted to let you guys know what's going on. And in a few weeks, it's going to be obvious anyway."

"He'll be a killer dad," Saher said to Lisa out of Amir's earshot. "I think he'll be fun and cool, not like a lot of paranoid, older parents."

"No," Lisa said. "But he *will* be old when his kid's a teenager." They giggled. "Well, not *that* old. Maybe semi-old."

With Amir enthused about the grant and the students cheered by his baby announcement, the team attacked the day's work. They started by writing a list of things they had to do on the large whiteboard hanging on the wall: Helix, Structure That Holds Helix, Turreted Shooter, Electrical, Wheel-Drive Mechanism.

It was a long list for so little time.

"Machine. Machine. Machine."

Throughout weeks three and four, as the team fabricated parts for the wheel modules, this was their mantra. Each module had forty individual parts, and the students needed to machine enough parts for eight modules (four each for the competition and practice robots). This was the burden of building an omnidirectional drive train. By contrast, a standard tank drive would have needed less than a quarter of the parts.

Amir often needed to spend several hours each day in the shop, showing his students how to machine individual parts. A line usually formed outside the shop with students needing his help elsewhere.

When Luke interrupted him during one session with a question about his wheel-module design, Amir held up his hand. "Give me a minute. First, I'm going to teach people how to fish, then I'm going to come fishing with you, okay?"

"I see." Luke smiled, already reaching for his notebook to add the quote to his list of Amir-isms. Every year one of his students kept such a list.

One afternoon, Amir instructed Turk and Alejandro on how to use the grinder to round off the aluminum rectangles they had cut for a module's side plates. The three put on bright yellow earmuffs and donned safety glasses. Amir went first, bringing the plate against the grinder's spinning belt. Even with the students wearing earmuffs, the screeching produced by the machine was excruciating. "Don't push hard!" Amir shouted. "If you do, you'll ruin the belt."

He showed them the various angled strokes against the belt that produced the best result, then asked, "Who's ready to step up?"

Alejandro stepped forward before Turk could. He took a few tentative strokes, getting a feel for how forcefully he needed to push. A minute later, Alejandro pulled down the earmuffs so that they hung from his neck.

"What, you like the noise?" Amir asked.

"This way I can hear how hard I'm pushing."

Amir just shook his head.

Most of the parts for the modules required absolute precision to machine, and some were intricately complicated. Students spent hours on the mill or lathe shaving off thousandths of inches of aluminum again and again to get the exact measurement needed for two parts to fit together. It was a level of precision that would have impressed a professional engineer. However, some students grew impatient at the slow, methodical process.

"It's so redundant in there," one student complained to Stan after emerging from the shop.

"You mean repetitive," Stan said.

"Yes, repetitive." The student shook off from his shoes the sparkly slivers of aluminum that now littered half the build-room floor.

"Repetitive is good. When you're doing the same thing over and over and you're doing it right, that's good news."

Throughout these days of intense machining, the noise from the shop was overwhelming. For the programmers and the SolidSeven trying to concentrate on their computers, the whine of metal cutting into metal was crippling.

Eventually, Max went to the shop where the Ladies of the Lathe, Saher and Lisa, were working on a hexagonal axle. He grabbed two pairs of earmuffs from the wall, and he and Colin Ristig wore them at their workstations while they fiddled with their designs.

At one point, the two called into the machine shop to see if their teammates could quiet it down. The hours standing at the mill and lathe had frayed nerves, and some of the old bad blood among students resurfaced in an instant. One student in the shop came out, face red. He gave them the finger and shouted a string of curses. Then he stopped, turned, and returned as if nothing had happened. Everybody was taken aback but continued working.

The lack of sleep was also causing a lack of focus.

One night, two slaphappy students were messing around while operating the mill, talking in high-class British accents. Amir came in, checked out the part they were cutting, and said, "You didn't do it right because you're too busy speaking in British accents instead of concentrating." The two looked at each other, not sure if he was kidding. He wasn't.

On another night, Chase mixed up an online order for a shipment of parts, setting them back a day. He told Amir, "I'm probably not the best guy to be typing things in quickly." Later that same evening, Amir took one of the vertical axles that Saher and Lisa had made and started cutting an octagon instead of a hexagon, ruining four hours of work. He had to stay even later than usual to fix it.

But nothing caused as many errors as inexperience. At one point they couldn't get the bevel gears they had machined for the wheel module to fit together. This provoked a long debate between Amir and Luke, who were trying to figure out what went wrong. At the start, Turk took dollar bets on who would win.

"If you don't speak SolidWorks," Luke said halfway through their back-and-forth, "then it's difficult for me to communicate it to you."

"How can you *not* speak in reality?" Amir sat down.

"Because I'm not *in* reality." Luke pointed to his SolidWorks design of the wheel module on the screen. His beret tilted awkwardly on his head. "I'm in *here*."

Amir also pointed at the screen. "This *is* reality."

Turk interjected, "He's in the virtual world."

"You're telling me that a distance changed between this surface and this surface," Amir said, gesturing between the gear on the screen and the one in his hand. "And you can't tell me why. That's what's happening right now."

"I've been trying to tell you why, but you won't believe me."

Turk patted Luke on the back. "You tell him."

Luke pointed to a measurement of the gear on the screen. "That's the distance between the back and the middle. That's how you define where the mesh is. It changed on the gear we made."

Amir explained that Luke needed to input values in Solid-Works that corresponded to the dimensions of the gears they had manufactured rather than the ideal dimensions (even if the difference was ever so slight).

Luke grinned. "That's what I did wrong."

"Okay, that's why we ended up in this conundrum of insanity. We're done." Amir stood, walked a few feet away, and turned back to Turk. "Who won?"

"Stalemate," Turk said, barely able to control his laughter.

"Stalemate? Are you kidding me?"

When Amir was busy with other tasks, Chase stepped in to manage any problems with the wheel modules. He also machined parts himself, ordered materials, and organized what needed to be made and when. He hovered over the freshly made parts— plates, axles, pulleys, and bushings—like a mother hen protecting her chicks. When a measurement was off, he had to make sure it was right. When the milling machine needed a different attachment for an unusual cut, he put it on, recruiting others to help. That was not to say he was all business.

"The Incredible Hulk's powers suck," one student said, furthering yet another superhero conversation Chase had instigated by praising the abilities of the X-Men's Wolverine.

"You have to get him mad first for him to use them," Luke added. "The only thing cool about the Hulk is his magic pants— no matter how big he gets, his pants don't rip."

Turk chimed in. "I guess that proves that the Hulk doesn't get mad *everywhere*."

Over their laughter, Chase said, "Okay, that's all. We've gone from X-Men to what's going on in Hulk's pants. Game over."

Despite these interludes, parts began to stack up at his workstation, an encouraging sign of progress. Chase was eager to start assembling the first modules.

The team was not making every single part, however. The chassis that Colin and Stan finished on Sunday at the Bat Cave was designed in SolidWorks so that it could be cut from a single sheet of aluminum. Then it would be bent along exact lines to create a short rectangular box that the wheel modules would fit inside. Valley Precision Products, a local machine shop whose owner Amir had known in high school, was a sponsor of the team and had volunteered to do this work. On Tuesday, January 27, the team traveled over to Valley Precision to watch the chassis for their practice robot get made.

On their return to Dos Pueblos, the programmers were abuzz.

Gabe started, "They program everything into this huge machine, put a piece of metal in there, then it moves it around—it has about eleven tools in it—and it cuts the whole thing up. It goes fast."

"It was like, *gumph, gumph, gumph, gumph, gumph, gumph, gumph,*" Kevin said.

"It's crazy cool to watch," Nick added, more animated than usual.

"I don't even know how they got that thing in the door," Gabe said. "They must have built the whole place around that machine."

Their excitement aside, the shrill whine of wheel-module parts still being fabricated reminded them that they still had a lot to do before they had an operable drive base.

By Thursday, January 29, the team was losing the intensity they had earlier in the week. With nineteen days until ship, Amir stood before his seniors in their second-period class to remind them of their dwindling number of build days.

"I want you to use every minute like it's precious. Work together the best you can, to move everything forward without my guidance. I know what I'm asking you to do is something you're not able to do yet because you don't have the skills, but you're close. I know it's difficult, and the fact that it's difficult is what's making it so you don't take that leap, but you have to take the leap."

Once again, the team rallied. So many students were trying to make parts in the machine shop that they created a bottleneck. Andrew Hsu took over the machining as Chase started to assemble the wheel modules. Andrew, who had a frame large enough to play football, had always hoped for a career as a cellist. But the more time he spent in the machine shop, the more he fell for what Amir described as "the primal need to build things with one's hands."

With no space in the machine room, another group of students went off on their own to make the bumpers in the school's wood shop. Within an hour, sawdust covered them completely.

The programmers renewed their focus on finishing the code for the turreted shooter.

By midafternoon, Amir told his team, "You're on fire."

Sequestered in a computer lab next to the build room, Max tried to finalize the helix. After they had decided to go forward with this concept at the end of week two, Amir sat down with Max and the others in his group, including Turk and John, to hash out its precise dimensions. He spoke of pitches, diameters, and right triangles. He explained that they needed to use the Pythagorean theorem to figure out the measurements of the sheet-metal cuts they would need to form the helix.

Turk and John nodded, as if the explanation was clear, but it seemed that they were just trying to let Amir and themselves off the hook. Max, who understood the design of the helix better than any of them, said he was still confused.

"You're smart," Amir had said to Max at the start of the third week. "You have these guys around you who are all smart. Tell them the problem, explain the geometry, and see if they can help you come up with the math to make it work. I'm asking you guys to be self-sufficient. This is like if you were a professional engineer. The senior engineer would ask you, the junior engineers, to do something and you do it. Or else that senior engineer will have to sit down and do it for you, and you *don't* want that."

Slowly over that week they figured it out together. If they took a single rotation of the helix and flattened it out, the result looked like a donut with a small slice removed from it. They would cut this flat pattern from a piece of sheet metal to form one layer of the helix. They would need to fabricate four of these layers and connect them to create the entire helix. When expanded, this helix would look like four turns of a giant Slinky.

They found that the helix climbed up the inside of the cylinder at a constant rate and they were able to relate this to the right triangle Amir had spoken about. Then, using their knowledge of angles, arc lengths, and circumferences, along with the Pythagorean theorem, they determined the exact dimensions of each layer of the helix to be cut from sheet metal.

Now in the fourth week, Max was finalizing the helix's dimensions as well as the designs of the other elements of the robot core. This included the rotating interior cylinder, the fixed outer cylinder, and the core's support structure. This structure was four posts fitted with four aluminum shelves spaced evenly from top to bottom. These shelves had large circular cutouts in their middle that fit the circumference of the outer cylinder. The top two shelves also had truss supports, similar to those on a bridge, to help carry the weight of the turreted shooter that would be placed on top of the robot core. The challenge in designing all of these elements was to make them lightweight.

That Thursday night, Stan, who had been helping Max, ran through his designs one last time, then said, "I'll take the drawings to Valley Precision tomorrow."

On Friday the team discovered that their bearing blocks, one of the key components on the wheel modules, had been made incorrectly. Fabricating new ones would take two days at least. Instead, Chase and Amir worked for half a day modifying the bevel gears to adjust for the mistake. Although these modules were for the practice robot, and they would know to avoid this problem when building the competition robot, they had still lost time.

"I said everything's going to be fine, and it's fine," Luke said, always the optimist.

"Except we lost about five hundred man-hours," Amir said.

Luke reached for his notebook of Amir-isms. "I got to write that down."

Across the room Gabe called out, "Five hundred man-hours. How many man-*seconds* is that?"

"Shut up," Amir said. "I only lose chimp-hours from the programming department."

"In freshman physics, you should have them do man-hour calculations so they can get ready for robotics," Gabe said, struggling to complete his thought without bursting into laughter. 'Thirty people are sitting in a room for an hour, and four people aren't working. How many man-hours are wasted?'"

"You. Guys. Are. Idiots."

Gabe almost fell from his chair, he was laughing so hard.

The team only had to look at the whiteboard to be brought straight back to reality. Chase checked through other parts made for the drive train and discovered several machining errors. Throughout the rest of the afternoon, Amir and the students rushed about the build room and machine shop to correct these problems so they could finish assembling at least one wheel module.

When John Kim arrived with the chassis machined by Valley Precision, the first big part of the practice robot, the students barely reacted. Gabe weakly said, "Yeah! Robot!" and there were a few mercy "Woo-hoos," but that was it.

Stan walked around the build room in a fog, his eyes puffy and red. Like the students, he was there every afternoon and Saturdays until late at night. He then spent every Sunday with one of the SolidSeven at his house. Of the mood in the room, he said, "It's a combination of mistakes being made and having a ton of stuff to do."

Despite this, Amir left early, a rare break so he could take Emily out for her birthday dinner. The build room devolved into chaos. Two students who tended to slack off even when Amir was there started making "ninja throwing stars" out of paper. Others huddled around a computer playing the video game Worms. Stan was the only mentor still there, and he was not one to scold the students back to work.

At eight-fifteen the team started to disband, the earliest they had gone home in weeks. Chase was reluctant to leave, even though he was already an hour late for a date with his girlfriend, a student at Santa Barbara High School. They had met right before the season, and except for Sundays they never saw each other anyway. "Bad timing," he said with a shrug.

As he walked out, several teammates started chasing one another around the build room on their kick scooters. Unless the team proved efficient over the weekend—unlikely since Saturday was the winter formal dance—they would fall even further behind schedule.

And then the plague hit.

On Saturday, Amir awakened feeling like something was amiss. He didn't have a fever. No cough. No nausea. It was more an uneasiness. Four weeks had passed since they had driven down to Los Angeles for the kickoff. The days and nights had been long, and the cumulative lack of sleep was catching up with him. But this was different.

He put it out of his mind and readied to leave for Dos Pueblos. Just before he walked out the front door, Stan called. "I'd like to meet face-to-face," he said. No explanation.

Amir hesitated for a second. He had always feared this moment. Stan would never say anything until one day he would just reach his breaking point and want to quit. From the sound of his voice now, that day had come.

If Stan left the team, Amir didn't know what he would do. They were already behind. At the start of the season, Amir had wanted to have a functioning practice robot by the beginning of the fifth week. They didn't even have an operable robot base yet. Without Stan to help him, they would be so deep in the hole that Amir feared they might not be able to climb out.

"I can't meet at my house right now," Amir said. "We can meet at school."

"Okay, fine," Stan said. Conversation over.

A half hour later, the two met in the build room and then headed to the teachers' lounge where they could talk in private. On the short walk, they exchanged very few words. In the lounge, they sat down at a table. Amir's earlier uneasiness had turned into a general feeling of weakness. He was also experiencing occasional waves of dizziness. His nerves over what Stan was about to tell him did not help.

For a few seconds, they stared at each other. Then Stan slid a piece of paper across the table. It was a timeline detailing what they still needed to do and the time it would take to do it. "I've gone through this," he said. "We won't finish. We need to cut some things out."

Amir studied the time line. Stan had written down what they

needed to consider eliminating in order to finish on time: re-tractable hood, rotating turret, the camera tracking, and the large number of balls they needed to store in the helix. If they wanted to ship a robot, Stan said, at least two of these items had to go.

It was a reasonable list, and Amir feared that if he didn't agree to eliminate some mechanisms, Stan might quit on the spot. Regardless, Amir knew he couldn't go down that road.

"I'm doing this with the students," Amir said. "We came to an agreement on the kind of robot we wanted to build. I can't go to them and look them straight in the eye and say it'll be okay to cut these things. I can't sell this to them. It's their project. It won't be the same robot they signed up to build. I wouldn't be able to live with myself."

Stan was speechless.

To finish in time, Amir said, they would "burn themselves into the ground," not sleep if necessary. He was also confident that once they had finished building the practice robot, the competition one would come together much more quickly.

"What about debugging?" Stan asked. They needed time to test and fine-tune their robot.

Amir said that they could do these tests after the six-week build period. The competition robot had to be finished and shipped, but *FIRST* allowed teams to keep their controllers after the deadline to refine their software and upgrade a limited amount of their mechanical systems. Since their team was building a practice robot that was almost identical to their competition robot, they could debug after the ship date. Any adjustments needed on their competition robot could be made in the eight-hour window that *FIRST* gave teams to work on their robot on the first day of their first regional competition.

"I see," Stan said, not entirely convinced, but willing to go along with the idea. However, he said that he couldn't keep putting in so many hours and maintain his business and his family life.

Amir said he understood and was just glad that Stan was still with him. As they walked back to the build room, Amir started to feel worse and worse: his hands were clammy, his stomach unsettled, and his legs wobbly. When he reached the build room, only half his team was inside. "It's an epidemic," a student explained. "Everyone's getting sick."

Oh no, Amir thought. He got on the phone with Luke, who had reported the night before that he was feeling unwell. "Describe the symptoms," Amir said.

He felt funky. Then dizzy. Then way more dizzy and nauseous. Then shakes and shivers and chills. Then fever. Then vomiting. Lots of vomiting. Then feeling slightly better. Then the cycle again, but worse.

Amir walked outside and felt like someone had punched him in the kidneys. He needed to sit down.

"Good-bye," Stan said, standing at his car, about to head home to look over some SolidWorks designs.

"In the interest of full disclosure," Amir said, "I'm about to die."

Stan wanted to know if there was anything he could do, but Amir was already retreating to his car, parked in front of the build-room door. He collapsed into the driver's seat, his students surrounding the car, not sure what to do. As Amir reclined his seat, he asked his ever-present student aide to do something about the sun shining into his eyes. John hurried back to the room, grabbed a sheet of regolith, and propped it onto the hood to deflect the sun. Another student ran to get Amir a jug of water and a trash can for vomiting.

Amir thought he'd suffer through whatever this sickness was in his car, using it as mission control. He couldn't leave now.

When Chase came to his window, Amir started to give him instructions on how to put a certain attachment on the mill to make the final parts for the wheel module. Chase didn't know what this collet looked like, but he ran into the machine shop, searching around, asking, "Is this it?" to anyone who would listen, until he found a green vise that he had seen Amir use before. That must be it. When he had the vise installed, he returned to the car.

Amir had deteriorated even further. He was shaking horribly. He tried to give more instructions, but his students stopped him. "You're still worried about the robot?" they asked, looking at him like he was mad.

Before losing it, he called Emily and asked her to come pick him up and take him home. He couldn't drive. Amir gave Chase a long list of what the team needed to accomplish that day.

"We can't screw up any of these parts. You're going to have to describe what I'm talking about to the others. Can you do that?"

"Yes," Chase said.

Emily arrived wearing an oversized T-shirt and pajama bottoms with pink and red hearts on them. She smiled. "He has a real flair for the dramatic, doesn't he?" she said before driving off with her husband.

Inside the build room, Colin, who had not surfed in weeks because of his long hours working with Chase on the drive train, stood before the whiteboard. He took a pen and wrote "The Plague," then listed Luke, Turk, Gabe, and three other students. As each hour passed, more students disappeared from the build room, and Colin added more names to the list.

"I want to hang out with winter-formal people tonight," Andrew, one of the survivors, said, taking a break from the machine shop.

As always, he was wearing a brown hoodie and seemed like he was speaking on a slow track.

"Me too," Colin said, tired of staring at his SolidWorks chassis design.

"I guess I can skip it if there's stuff we can do here tonight."

"I can stay too."

"That's tight," Andrew said, his stock phrase for agreement.

"This is such a crucial time," said Kevin, who had put in almost as many hours programming as Gabe. "If we don't keep pushing forward, we're screwed."

Chase led the charge, giving everyone a list of parts they needed to make. When one of his teammates' fathers, a UCSB physics professor, arrived to help the team out, Chase even gave him his marching orders. Initially Chase worried that nobody would listen to him with Amir absent. When this worry passed, he thought only of staying healthy and not making any mistakes.

Still, at the end of the day, they were even further behind than before, and with Amir sticking to their ambitious plans for the robot, the team ran the risk of not finishing. They were in big trouble.

But trouble was relative. For some *FIRST* teams, participation in the robotics competition might mean the difference between going to college or a life abandoned to the streets.

The Competition

2TRAIN ROBOTICS, TEAM 395

On Sunday morning, January 4, Gabe Ruiz stepped out of his South Bronx apartment building, the wire-meshed security door clanging shut behind him. Crossing the street, he slipped in the earbuds from his Sony PlayStation Portable (PSP) and then pulled down his black skullcap. He figured he was safe on his walk past the projects to the subway. With the cold and early hour, the thugs wouldn't be out. They were lazy that way.

By the time Ruiz reached 163rd Street, the thrashing metal music of Slipknot pulsed in his ears. The lyrics from his favorite song, "Vermillion," repeated again and again: "I won't let this build up inside of me." He felt like the words were written for him, for the anger that he felt. He hated being at home. He was sharing a small bedroom with his brother, who was more than twice his age and just back from prison. His adoptive mother, who was sixty-nine years old, very religious, and only spoke Spanish, was a stranger to both his struggles and aspirations. When asked about the differences between Ruiz and his ex-con brother, she recounted none.

School was no better. Having to pass through a metal detector at the school entrance every morning made him feel like a criminal. His classes bored him. His grades were bad, and he had one semester of high school left. There were a lot of classes he was struggling to pass just to graduate. After that, he had no idea. The only thing Ruiz knew for sure was that he wanted out of his neighborhood, where drugs and gangs—and plenty of stupidity—ruled.

In the past two years, he had been unable to escape its violence. After his sophomore year, he felt like he had crossed some Rubicon where the street found it all right to come after him. Time and again he had guns and knives pulled on him. On Halloween night his junior year, Ruiz and a pair of friends had been cutting across a playground on the way home when a gang started throwing eggs at them before attacking. Ruiz landed one punch before several guys knocked him against a car and kept kicking and punching him until they tired. Ruiz still had a lump behind his ear from the incident.

After that, he had stopped caring much about school. He no longer listened in class or did his homework. His grades dropped from As to Cs. For a while his teachers kept pushing him, but most of them gave up after a few months. He retreated into reading books, listening to music, training for fights in mixed martial arts, and carving portraits into blocks of linoleum with a knife. Now he felt like he had fallen so far behind at school that he wasn't sure if he could ever catch up. Worse, sometimes he thought it might be better not to make the effort at all than to fail trying.

Once Ruiz reached the 2 subway line, he dug out from his backpack the book *Under the Wolf, Under the Dog*. He saw a lot of himself in the narrator, a sixteen-year-old sent to a treatment center for troubled teens, who writes in his first journal entry: "By the time anyone reads this, hopefully I'll be out of this place and on to better things." An hour later, after switching trains, Ruiz walked up out of the subway, back into the cold. The projects were gone now, and he stood in front of the wrought-iron gates of Columbia University in upper Manhattan. Already he felt better. There was nobody to watch out for. He'd worry about school and his family a different day.

He crossed the Ivy League campus, walking alongside several college students. He headed up the steps and around Low Memorial Library. Its tall columned entrance and monumental granite dome still awed him as much as it had the year before, his first on 2Train Robotics (Team 395). He walked north toward Seeley Mudd Hall, a brown rectangular building that lived up to its nickname, "the Brick."

After an elevator ride down two floors and a short walk in a barren white hallway, he arrived at the Columbia Mechanical Lab, room 294. Before he even walked through the doorway for the first day of the build season, he smelled the metal and oil. He felt like he had come home.

The students of the 2Train Robotics team did not plan on building the most complex robot they could imagine. They knew that winning the championship was more than unlikely. Their mentors did not set their sights on revolutionizing education or inspiring future leaders of industry and engineering. They would consider their season a monumental success if every kid graduated high school and pursued some kind of advanced education.

Most of 2Train's students attended the Morris Academy for Collaborative Studies, a public school whose student body was primarily black and Hispanic. Many of the students came from single-parent homes on the verge of poverty. Housed in a castle of gray stone and brick in the South Bronx, Morris was labeled a level-one high school. As its principal explained, this essentially meant that its students had been victims of educational neglect long before they ever stepped through the school's doors as freshmen.

A Morris teacher with spunk but no engineering experience had started the robotics team a decade before and had recruited Bob Stark, the lab manager for Columbia University's mechanical engineering program, to mentor the kids. Bob convinced his school to allow the team to build their robot in Columbia's advanced, multimillion-dollar machine shops, where he worked. University students signed up to help. In 2006, two freshmen from an elite private high school joined as well. One of these students, Adam Cohen, was the great-grandson of Adolph Ochs, the man who built the *New York Times* into an institution. When Adam had tried to start a team at his own school, the headmaster told him, "Robotics is a fad middle school boys grow out of. You should play soccer."

At the heart of this disparate medley of a team stood Gabe Ruiz. Smart, imaginative, capable with his hands, well-meaning, and spirited, he could exemplify the best in all of them. Brooding and self-destructive, he could also embody the worst.

"My idea is a conveyor belt that runs along the top," Ruiz said, standing at the lab chalkboard. He was presenting his design for their robot in an early brainstorming session that first week of the build season. "When we want to throw the balls out, maybe there's something on the conveyor belt, something to grab and drop them into the trailer." He stuck out his jaw, the muscles in his neck tight, as if he was waiting for a punch.

"That's inventive," Bob said. He Rollerbladed to and from Columbia every day, and his salt-and-pepper hair and the crow's-feet around his eyes were the only signs that "the Bob" was fifty years old.

"Good work," Adam said, wearing one of his many robot-themed T-shirts. This one featured two machines about to embrace. The line underneath read ROBOTS HAVE FEELINGS.

"Thanks, man," Ruiz said, smiling.

After sketching their ideas on the blackboard, the team settled on their design: a simple four-wheel tank drive with a vertical tower that delivered balls to a fixed shooter. It was not a robot that would inspire awe at competition, but it was one they could build.

Ruiz liked the sanctuary of the lab. He liked feeling useful. He liked working with the machines. He liked the responsibility of watching over the freshmen to make sure they safely operated the jigsaw. He liked assembling the drive train with Adam, seeing himself as his equal. The problem was his time outside the lab.

"Hey, get over here!"

At the start of the third week of the build season, Ruiz was walking toward the subway when the three thugs waiting at the corner must have seen him pull out his PSP. Ruiz looked up at them, calculating the danger and his options in an instant. That was all the time he had.

"I don't want no trouble," Ruiz said. He placed his hands behind his back like he had a gun while stepping toward them.

The thugs each took a step backward, and one of them placed his hand behind him at his waistband as well.

There was a long, soul-sucking pause.

"Let's go," the thugs said. They backed away. Ruiz backed away. It was over. For the next hour, Ruiz walked about the streets to work off the bad rush of adrenaline and the feeling that he was about to break apart if he didn't escape this life of his in the South Bronx.

He arrived at Columbia at six o'clock in the evening and headed straight into the machine shop, where Bob was cutting pieces of ¾-inch aluminum tubing for the robot's frame.

Ruiz and Bob remained in the machine shop until ten o'clock, long after everyone else had left. In his mad-professor blue lab coat stained with spots of oil, Bob showed Ruiz how to use one of the machines, but mostly they just talked. For a long time, they debated the merits of various slasher movies and their favorite reality shows. Ruiz liked the show *Ghost Hunters.* "Makes me feel there's more to just being alive," he said. In the spaces between these conversations, Bob told him he had the potential to do anything he wanted in life, starting with going to college. Ruiz liked hearing the words. The trick was believing them.

The day after his talk with Bob, Ruiz decided he should try to go to college. Enough shutting everything out of his life so that he wasn't vulnerable. He had been doing that since he had been jumped on Halloween, thinking of himself as a ball—with no exposed edges, impenetrable—rolling through his life. But he knew he was just sleeping through his days. He needed to make a change.

At Morris, he filed his applications for several New York State colleges. He felt good about it but little relief. He still had his state exams and finals to pass to graduate. More frightening, he was now exposed to the potential of failure, and he might let down those few on the team who believed in him. After he signed his name to the college forms, these fears overwhelmed him.

He skipped robotics the next day and started coming less and less. By February 1, he had gone a week straight without attending. The rest of the team was hit-or-miss at best. Those who did show trickled in later and left earlier each day. In their few hours there, they were a listless bunch. "Why aren't the kids interested anymore?" one of the Columbia student-mentors asked Bob.

Everyone missed Ruiz. Finally, the mentors tracked him down at school, urged him to return, and told him he was important to the team. "Of all the kids I've had," one of them told him, "you're at the top. Do you know how many people care about you and want to help you?" He told Ruiz they were not going to let their expectations of him slide away into disappointment.

Ruiz returned. Team spirits rose, no matter that the tower they had built to deliver balls to the shooter was inoperable. Without it, theirs would be just a defensive robot. They soldiered on. Ruiz came every day.

There was no sudden Hollywood turnaround. Ruiz did not start acing every class, score perfect on his college-entrance exams, and somehow get a full ride to Harvard. But a community college in Upstate New York accepted him, and he was planning on attending this school over the others he had applied to because it was the farthest away from the South Bronx.

After an overnight build session right before the ship date, the team would finish their robot. It was a collection of aluminum shafts, plywood, chains, sprockets, motors, cardboard tubing, electrical wire, rivets, and even dental floss, the sum of which looked like a mailbox on wheels. The tower worked now, some- what. It was theirs though. They had built a machine that at the start of the season they had not imagined possible.

"Kick the tires and light the fires," Ruiz declared after lifting the robot into the crate. "I'm ready to roll and take home the gold."

When one stripped everything else away, the story of Ruiz and his 2Train teammates was about expectations. Few, if any, of them had ever had much expected of them. Now, to be part of the robotics team, the Morris students were expected to travel to Columbia—an hour each way every day after school and on the weekends—for meetings six days a week. They were expected to maintain good grades, help come up with ideas, machine parts, wire the electrical board, learn programming, study in their downtime, respect one another and their mentors, and take everything they were learning and imagine a future beyond the South Bronx.

With more than sixteen hundred teams participating in almost fifty regionals across the country, Amir would never know the story of Ruiz or 2Train. But the ability of this robotics competition to raise the bar on what was expected of high school students, no matter their background, was the reason Amir centered his capstone course on *FIRST*. And if he won his grant, he aimed to recruit disadvantaged kids like Ruiz to be a part of his expanded academy.

Week Five

15 DAYS UNTIL ROBOT SHIP

He's going to break me.

—KEVIN WOJCIK

It was Tuesday, February 3, and the Dos Pueblos build room was a shambles. Aluminum shavings and plastic zip ties crunched underfoot. Open boxes of parts stacked haphazardly in the corners looked ready to topple. Swimming-pool noodles to be used in the bumpers were scattered about the floor. A flat rectangular cardboard package containing a sheet of thin plastic lay in the center of the room; everybody navigated around it and nobody seemed to want to pick it up. Parts of the helix for the practice robot, fresh from the CNC machine at Valley Precision, were leaning against a chair. Empty Slurpee cups, soda cans, candy wrappers, and chip bags littered workstations. Tools were everywhere except in the boxes where they belonged. The air smelled of thirty-one high school students who had spent way too much time closeted in the room.

The team had survived the plague. As students returned to the build room, they erased their names from the list on the whiteboard. They limited their conversations to either how far behind they had fallen or what had caused the sickness that had helped put them there in the first place. Some conjectured that it had to be something they ate at their group dinner. Those who had been stricken remained hollow-eyed and sluggish, but at least they were there.

Chase had escaped the bug and spent all day Sunday and most of Monday in the build room, trying to keep things on track. He now sat at his workstation, putting together the wheel modules. He had gone into the season thinking that his abilities and experience would make him the most valuable member of the team. Now he thought differently. Without the SolidSeven, the robot never would have been designed so well. Without Andrew, Saher, Lisa, and so many others on the team, the wheel modules never would have been machined. He understood now that everybody was needed, that he was only as good as how well he worked with the others and how many hours he put in beside them.

Still, he was the right person to be in charge of assembling the omnidirectional drive train. On his workstation, he had spread out the forty parts of the modules for the practice robot. They looked like the pieces of an impossibly complicated puzzle. Chase knew where and in what order they fit together without needing much more than a glance at the printed-out design. He put the first module together starting with a kit wheel; next came an axle, then a pulley, and so on. When a spacer between the side plate and gear was not flush, he sent it back to the machine shop for a shave. When a timing belt felt loose, he realigned the pulleys until they were straight. When he finally secured the last of the two pulleys to the wheel module, he turned the top one to make sure that its movement translated into rotating the wheel, then he turned the bottom one to check that the entire module pivoted 360 degrees.

Finished with the first module, Chase set out to assemble the second. After he finished that one, he would secure the two modules to a long rectangular supporting plate and arrange the timing belts and pulleys that connected the two modules to form half of the drive train. The chassis machined by Valley Precision now rested on the floor, an empty shell until they set the wheel modules within it.

Colin, Luke, Saher, and Lisa were by Chase's side. They worked together almost like they were on an assembly line. Next to them, Max, Turk, and a pair of other students began to construct the support structure that would house the helix. Amir liked what he saw as he walked through the room. "Finally, the factory environment that'll get us done," he said.

As he strode away, Stan whispered to several students, "This right here's why you want a quality education. You don't want to work in a factory."

Just as everybody started to feel enlivened, they were reminded of the toll the build season had taken on them. After dinner that night, in the middle of the room, John was resting facedown on a big exercise ball, his arms wrapped around it in a bear hug. Amir's perpetually tired student aide was perfectly balanced— a nudge to his side would have sent him toppling over. Several of his teammates noticed that he had been in the same position on the ball for more than a half hour. A crowd formed around him, and quiet whispers of "Is he asleep?" became "No way he's asleep" to "Dude, he's totally asleep." They left him alone, everybody as exhausted as he was. About fifteen minutes later, there was a crash. John had fallen and knocked over several boxes along with a drill.

Everybody hurried over to see if he was okay.

"You were sleeping on the ball, do you realize that?" Amir asked, helping him to his feet.

"No, I was working on the lathe," John said in a total daze.

The room erupted. Amir reminded them of a talk he had given about a student the previous year who fell asleep while driving home from robotics. "Be aware of your limitations. I'll push you hard, but I don't want my temperament to put you in a bad situation. If you're too tired at three A.M. to drive home, call your parents."

Amir advised John to get more sleep, then he was off. The arrival of several boxes of cold pizza sidetracked the team again. The electrical crew warmed up the slices with a heat gun.

"Guys, listen up," Amir said later. "Realistically, I'm working with the drive team until the end of the night. We need to drive ASAP. Other people create a list of parts that need to be made. Programming people, go write some software."

"Your code doesn't work," Nick said.

"You mean *our* code doesn't work." Gabe clutched an AriZona iced tea, the mix of caffeine and sugar keeping his eyes from rolling into the back of his head.

"Okay, *our* code doesn't work."

At the start of the fifth week, the programmers were writing code for several different mechanisms of the robot at once.

"Did it crash?" Nick asked about some code they had just run on the drive train.

"No," Kevin said, typing on his laptop but looking at a large flat-screen monitor they had brought in to save their eyes.

"Are you sure?" Gabe asked. "I think it crashed."

Kevin reached for a chocolate bar in his backpack, which was resting on the workstation. He knocked a ball off the shelf, which hit first the monitor, then his laptop. Everything went dark. "Now it's crashed," he said.

Later, they were programming a kill switch for the robot. "We need to make sure that the switch stops the robot." Gabe entered a last bit of code. "This should work. How many times have I said, 'This should work'?"

"About 135,000 times," Nick said.

The test failed.

"Has this switch ever worked?" Gabe asked, referring to the actual switch on the robot operator console that Chase and Kevin would use in competition.

Nick turned toward Gabe, the blank look on his face saying everything: *Don't tell me this thing has never worked after we've spent valuable time writing code to test it.*

Confusion reigned whenever the programmers explained to their teammates what they were doing. Over the past few days, they had been trying to write code that would count the number of balls in the helix. They hoped to have an opaque outer cylinder so that their opponents wouldn't know when they were loaded to score.

"This time, we should end up at F," Gabe said about their prototype counter.

"Why?" Chase asked.

"We're counting in hex, it's base sixteen."

Luke called out: "Fire when ready."

Fifteen balls tumbled down a ramp. Gabe started muttering letters.

"Let me do it, since I can count in regular numbers," Chase said.

"But we can't put you in the robot," Luke said.

Gabe grumbled that the sensors only counted to eight. "It's like seven off from what it should be."

"This doesn't make sense, man," Chase said.

"We count in base ten," Gabe said. "One through ten. Computers count in Base 16. Zero through F. So it goes zero, one, two, three, four, five, six, seven, eight, nine, A, B, C, D, E, F."

"That's so stupid," another student said.

"Why doesn't one through ten work?" Chase asked.

"Because there are eight bits in a byte," Gabe said.

The conversation devolved from there.

"What if we did everything in hex?" Luke grinned.

"What if we did everything in strings?" Nick replied.

"We'd probably all hate ourselves," Gabe said.

The truth was that nobody cared how the programmers did what they did. The team only cared that it worked. The pressure was on them now to program the code for the drive train.

The team's robot pilots, Chase and Kevin, had decided how they wanted to control the omnidirectional drive system with the joysticks.

"It's Halo," they explained to Amir, who didn't find the explanation sufficient but trusted them.

Chase and Kevin had wasted many hours mastering Halo, the science-fiction video game where a hero armed to the teeth gunned down alien hordes to save humanity. The game was played with two joysticks. The left controlled the hero's movement in any direction. The right controlled his orientation, or the way he was facing so he could find and destroy his enemies on the battlefield.

Both students wanted their robot to operate the same way. The left joystick dictated their speed and movement in any direction. If Chase wanted to move diagonally forward and right at full speed, he shifted the joystick in that direction, all the way to its maximum position. While the robot maintained its orientation, facing straight ahead, all four omnidirectional wheels pivoted at a 45-degree angle and started spinning.

The right joystick controlled the robot's orientation. If Chase wanted to rotate the robot around on its axis until it was facing the opposite direction, he shifted this joystick to the left. The wheels remained in a straight position, but the right wheels spun forward, the left wheels backward.

Programming these controls was as complex as the mechanics in putting together the wheel modules.

Changing the robot's orientation with the right joystick was the simpler of the two. In this movement, the wheels were always pointed straight, so they only needed to worry about how much power they were giving the motors driving the wheels and whether they were spinning them forward or backward.

Controlling the omnidirectional drive with the left joystick proved infinitely harder. They had to translate the position of the joystick into both how fast they wanted the robot to go and in which direction. The distance the driver pushed the joystick away from its neutral position dictated how much power to give the motors driving the wheels. The angle he pushed the joystick away from its neutral position dictated the wheel direction.

There was a lot of math and plenty of missteps involved in figuring how to interpret the joystick movements into numbers and angles that the controller could use. But Gabe found the effort rewarding. "It always surprises me when math works out in real life. My theory has always been that trigonometry is just some magic that was made up to make math work."

Rotating the wheels to the desired angle was the next challenge. To start, their code had to know the direction the wheels were pointed in initially. This required placing a sensor called a potentiometer on the wheel module's steering motor. A potentiometer worked like the knob that controlled a stereo's volume. Instead of providing a decibel level by how much it was rotated, the sensor indicated a specific wheel direction. The potentiometer would constantly update the controller on the wheel's angle.

The controller needed to send precise signals to the robot about how much power to deliver to the motor rotating the wheel module. This power needed to be sent in a measured way so that it didn't jerk the wheel from one position to the other. In order to achieve this, Gabe had to program a PID (proportional-integral-derivative) loop. Dozens of times a second, the PID loop would read the potentiometer and calculate how far the wheels needed to move to reach their target position (proportional), how long they had been off target (integral), and how fast they were moving (derivative). The sum of these calculations was used to adjust the amount of power delivered to the motor that steered the module.

People make these adjustments instinctually when reaching for a steaming mug of coffee on a table. You move your hand quickly toward the mug, knowing it's far away. But the closer your hand is to the mug, the slower you move until you grasp the handle. The robot needed code and input from sensors to make these calculations.

Controlling the wheel position with a PID loop was the same way Gabe had programmed the turret to rotate. He was close, but he had not mastered either. Close wasn't good enough. When the loop wasn't tuned properly, the turret or wheels would either move sluggishly to their final position, oscillate wildly back and forth, never get to the final position, or not work at all.

Gabe felt overwhelmed. While everyone had a lot to do, Amir could always throw more bodies into working on the shooter or helix. With the programming, only three of them could accomplish what needed to be done. And Kevin and Nick knew that without Gabe, who was basically programming at the level of a professional software engineer, they were lost. Evidence of this was a note written by Kevin on the small whiteboard over their workstation: "I feel like we achieved something without Gabe. I'm pretty psyched."

Burdened by this responsibility, Gabe was clocking more hours than any of his teammates. He was always one of the last to leave the build room. Once at home, he often sat at his computer programming for a few more hours. His mom was now frequently away in Los Angeles with his father, so he was all alone. On occasion he would take a break at night, texting friends or watching a little television to clear his head. *Battlestar Galactica* was one of his favorite shows. But once the show was over, Gabe was back at it. Even sleeping, he seemed to work through programming ideas. Several times during the past few weeks he had awakened with a different approach and jotted down his idea in the notebook he kept by his bedside.

He had expected the build season to be tough. From his theater-production work, he was used to long sessions, sometimes extending late after school and on weekends before a big opening. But those were short, concentrated bursts of activity, nothing that consumed weeks.

The time spent focusing on the robot code proved a much-needed distraction from his worries about his father. Still, the concern was never far from his mind. At the beginning of the season, he had made it down to Los Angeles a couple of Sundays in a row. It hadn't been easy staying in the room. He had to wear a mask. His father looked weak and tired. The conversation was uncomfortable. They didn't know what to talk about.

Because of robotics, Gabe hadn't made the journey the previous two weekends. On Wednesday, his father would be finished with the bone-marrow transplant. He had weathered the worst of it but needed to remain in the hospital. Only now did Gabe know how close his father had come to dying in the past weeks. One night during his treatment, he had a fever of 105 and told his wife that his greatest regret was that he wouldn't be able to see Gabe go to college.

It was a lot for a seventeen-year-old to bear. A few minutes before midnight on Tuesday, Amir had to force Gabe to leave the build room. He and Chase were trying to secure a potentiometer on one of the wheel modules, but they couldn't get it to fit right. "I like your tenacity," Amir said. "But efficiency is dropping."

Still fired up, Gabe and Chase drove to Freebirds, a Mexican-food joint near the UCSB campus that was their favorite late-night haunt. While devouring burritos, they spoke of robots.

At 7:40 P.M. on Wednesday, Amir issued a challenge: He wanted the wheels on the practice-robot base powered and spinning by an arbitrary 8:04. The drive-train crew had already finished assembling the modules and connected all the pulleys and timing belts. The electrical crew, led by Angie Dai (the team's "weight source" in their earlier drive-train tests), had wired up the motors and sensors and placed the electrical board on top of a small square sheet of plywood set on the chassis.

The programmers were now in charge. Gabe sat on the edge of a workstation, punching the last bit of code into his laptop to get the wheels to move. Kevin was attempting to set up the wireless connection between the operator console with its joysticks and the robot. Nick was crouched beside the robot base, not sure what to do. Everyone else hovered around, waiting to see if they had a drive train that worked. Given that the ship deadline was thirteen days away and they were still building the practice robot, tensions ran high.

A few minutes passed as Gabe and Kevin struggled to get the wireless router to connect. Luke stretched his long, thin arms over his head and joked, "Geez, why do you guys suck so much?"

The programmers weren't laughing.

"Now connecting to the robot," Kevin said, checking his watch. "We have nineteen minutes to get this running."

On the electrical board, two small indicator lights started blinking. Something was wrong with the wireless connection.

"Wait!" Gabe said, waving his hands. "Undeploy. Turn it off."

"No," Nick groaned.

If they had not stopped the code, it might have made the drive train run out of control.

They reset everything and restarted the controller. "Okay, I'm ready," Kevin said. "Wait, it didn't spawn the tasks."

Amir heard the statement on coming out of the machine shop. He had no idea what his programmers were talking about and said, deadpan, "I hate when my robot doesn't spawn its tasks."

Three of the connector lights were now solid, confusing Gabe even more.

"Why can't we just use the basic software that ran the six-wheel drive train and not the complex code you think will work for omnidirectional?" Amir asked, a little testily. He had only slept two hours the night before and looked walleyed.

"We're not," Gabe said.

"This is the most basic code if I'm just trying to spin a motor?"

"Not really." Gabe looked at Amir. He could tell how stressed his teacher was by the state of his shorts. That night his pair was more ragged than usual.

"You're wasting our time. What's our goal right now?"

"To try to make the motors move."

Amir explained that they needed to know if the drive train worked and what modifications they had to make. Nothing else. "Please let's test that, then you can put any fancy code you want on—stay here till four in the morning, I don't care."

He walked away. Gabe muttered an expletive, then said, "We're so deep into this, we don't know how to do simple."

"He's going to break me," Kevin said of his teacher.

Amir came back a few minutes later, his tone softened. "I just want us to be ready for the highest level of competition."

At 7:55 P.M., with nine minutes to go, Gabe loaded a basic set of instructions that spun the motors when the joystick was moved forward. All four lights indicated that the robot was ready. Kevin pushed the joystick forward. The wheels spun like mad. Timing belts flailed. The robot base shifted a bit forward on the carpet, then a belt connecting the two wheel modules on the right side of the chassis flung off. They cut power.

"That's not good," Chase said.

Amir had been sidetracked and missed the test. He wanted it run again. When he had trouble replacing the belt, Chase stepped in and did it himself.

One minute to go. This time Kevin ramped up power to the wheels slowly. All four began to spin perfectly, and the robot moved forward a few feet.

"Now gun the motor," Amir said.

Kevin pushed the joystick full forward, and a few seconds later, the belts flew off again. Still everyone cheered and started high-fiving one another. Gabe was more sober. "Other teams have finished their robot, and our drive train kind of moves."

They set the wheels on a sheet of regolith. Angie perched on top of the chassis to simulate the robot's weight, careful of the maze of wires running from the electrical board. This test would prove whether they had designed the modules with the right gear ratio. If it was too high and there was too much torque, the wheels would slip. Gabe had inserted more code into the controller's drive-train manager to allow the modules to steer left or right. This was a much more important test than the one before.

Just as they were about to start, the phone in the build room rang. It was for Amir. He walked away.

"Want me to floor it?" Kevin asked.

"No!" everyone yelled, Angie loudest of all.

Kevin pushed the joystick forward. The pulleys and belts began to turn, the wheels spun, and the robot base moved forward on the regolith. There were smiles everywhere.

"Do the commander move," Gabe said, referring to a strategy that Amir wanted to use with the omnidirectional drive where the robot would move in a semicircular path, the trailer stationary behind it. This would protect their trailer from being shot into by their opponents. Amir had borrowed the idea from basketball, when a guard keeps his opponent in front of him while protecting the basket behind.

The wheel modules pivoted as Kevin directed them with the joystick, and the robot performed a perfect arc.

When Angie stepped off the chassis, there was a huge round of applause, for both her and the robot. Amir was on the phone the whole time, his back to them.

"It's like a parent missing the baby's first steps," Luke said.

"No," Max said, his lisp more pronounced the more stressed he became. "That's what will happen to him during next robotics season."

Finally, Amir returned. "I can see you guys don't care about me. Can I see our robot now?"

With Chase on the controls for the first time, the drive train operated equally well. Although pleased, Amir and the team now focused more on what needed to be fixed. They figured they might need a lip on the edge of the pulley to keep the belts from slipping off.

"Shaeer, can we put our code on it now?" Kevin asked.

"Yes, you can put on your fancy code that doesn't work."

The next night, with this code loaded and new pulleys installed, the robot maneuvered around several sheets of regolith in any and every direction they wanted. It performed circles while maintaining its face-forward orientation. It moved diagonally, then sideways, then spun around its axis. It was also whisper quiet because they were using belts and pulleys instead of the chains and sprockets most teams employed on their robots.

"Is this thing sick or what?" Amir said. "It's almost as if it has a life of its own."

"Now can you start sleeping again?" Stan asked.

"Great job, everybody," Amir said.

Just for fun, they set up two boards on the regolith to create a space just wide enough for the robot. "Okay," Andrew said, stepping away. "Parallel park it."

Kevin rolled the robot forward until it was beside the open space. He turned the wheels 90 degrees, then slid right into the spot without a single adjustment.

"D-Train is pretty talented," Chase said, high-fiving Luke.

Colin, who had designed the chassis, walked up to the whiteboard where they kept their to-do list. He pushed his blond hair out of his eyes, grabbed a marker, and started writing. Nobody knew what he was doing until he turned to the side. On the board he had written this: "Successfully Make the Most Complex System This Team Has Ever Built."

"We can cross this one off," Colin said, drawing a line through the words with a flourish.

Kevin continued to drive the robot base around the regolith.

Amir added, "We have a long way to go, but if everything goes as smoothly as this, we'll be all right."

The Dos Pueblos team had much reason to celebrate, but as Amir said, they were far from the finish line. For one, they were still designing key elements of their turreted shooter.

In the past weeks, the crew of students working on the shooter had faced many challenges. Their makeshift wood prototype had come together in the first few days of the build season, but progress had been slow ever since.

Stuart Sherwin was the SolidSeven member in charge of the shooter design. He was now growing a goatee to complement the mane of hair that he hadn't cut in almost eighteen months. There was no doubt about his brilliance, though he carried it quietly. He earned straight As throughout high school, scored nearly perfect on his SATs, and planned on following his father, a UCSB professor, into physics. Whenever there was some controversy on the team over some random, esoteric subject (like how color blindness is passed from one generation to the next), Stuart always seemed to have the answer, delivered only if everybody else seemed stumped.

Once the prototype was finished, Amir had told Stuart and the others responsible for the turreted shooter to "get the basic geometry right. Dial that in, then worry about everything else." Figuring out the geometry took a couple of weeks.

The turret itself was a flat plate fitted with a lazy Susan that would be placed like a lid on top of the robot core. The key to the turret was integrating it with the core's helix. When balls reached the top of the spiral ramp, they needed to come up through a hole in the flat plate into the shooter hood. The shooter mounted above this hole needed to be in a position where it could rotate without sticking out over the side of the robot, violating *FIRST*'s restrictive dimension rules.

Then there was the shooter. To propel shots effectively, they had to design it with the right ball compression between the flywheel and the fixed hood. A second, adjustable hood would be placed over the first one to change the angle of the balls when they came out of the shooter. This hood would allow the robot to score on trailers at different distances. The shooter had to be compact but also provide enough space to mount the camera and other sensors for the vision-targeted tracking system that Gabe was programming.

"My brain hurts," Yidi Wang had said at this stage of the design when there were so many variables to resolve. Often she and Andrew Hsu took breaks to debate the merits of their favorite classical composers.

In SolidWorks, Stuart fiddled with scores of versions of the shooter until he managed to get all the parts to fit together properly. The hole was off-center by a few inches, and he decided to add a "kicker wheel" at the top of the helix that would force balls up into the shooter. After many tests with their prototype to find the best arc to sink balls into the trailers, Stuart designed the fixed hood so that balls came out at a 45-degree angle. These would be for the longest shots. The adjustable hood would stay clear of these shots when fully retracted and could also extend over the fixed hood to provide a range of trajectories. When it was in its maximum forward position, ideal for close-up shots, balls would come out at a steep downward angle into trailers.

"I like it," Amir had said. "The geometry's convincing me that we have it, or at least we're on the right planet."

Then Stuart had started working with Stan, whenever the mentor had free time away from the drive train or helix. Together they hammered out the details, mostly in week four and the early part of week five. The turret alone had twenty-eight parts, including the flat base plate, the lazy Susan, a motor and pulleys to rotate it, a potentiometer to measure its position, bearing blocks, brackets, spacers, and screws. The shooter had forty-four parts, including the two hoods, a pulley and timing-belt transmission, a flywheel, a camera mount, and two potentiometers. As with the wheel module, the design needed to be precise, and Stuart and Stan spent hours on SolidWorks moving mounting holes around by thousandths of an inch until they were ideally positioned.

They also spent time getting the shooter's aesthetics to match up with their mascot, the Penguin. The team name, the D'Penguineers, had originated several years before when Amir had been looking for a web address for the academy. One of his students joked that DPEngineering looked a lot like D'Penguineering, and they had their name. Each year the team called their academy's robot the PenguinBot and made it resemble a penguin in some way. This year the SolidSeven had designed most of the robot black, and Stuart was thinking that the hood could be gold like a penguin beak and the shooter wheel red to resemble a tongue.

By Thursday, February 5, while the drive-train team was celebrating its success, Stuart still needed to design the mechanism that would move the adjustable hood back and forth. He and Amir sat down for a two-hour concentrated stretch. The motorized mechanism they came up with to adjust the hood angle looked like a piston that moved up and down.

Amir drew a part on a sheet of paper. Stuart designed the part in SolidWorks. Amir adjusted the dimensions to his best guess of what was needed. Then Stuart printed out the design and placed it in a Ziploc bag. Their machine-shop master, Andrew, then came by to pick up the bag and went to make the part they needed. In that two-hour period, they designed and produced almost every part for the mechanism. With twelve days until robot ship, they needed this kind of efficiency or they were lost.

"Midnight is the new nine o'clock," Amir said, suspecting that their build session on Friday, February 6, would go late. They were set on finding out if the robot core that Max had designed to deliver balls from the intake roller to the shooter worked. First, though, they had to assemble all the parts of the core, including the helix, as well as the spinning inner cylinder and fixed outer cylinder between which the balls would be compressed and propelled up the ramp.

In the middle of the build room, Max sat on the floor beside the helix's metal frame. He was drawing diagonal lines with a black dry-erase marker on a very thin, flat sheet of Lexan that would become the outer cylinder. Often used in robotics, Lexan is a tough but flexible plastic. Lightweight, resistant to cracking, and transparent, the material is employed in bulletproof windows, like the ones around cashiers in convenience stores.

The Lexan would be rolled into a cylinder that would form the helix's outer core. The lines Max was drawing would provide a precise guide for where they would place the spiral ramps of the helix. The pitch and distance between each successive spiral needed to match up with what Max designed in SolidWorks, or the helix wouldn't fit right. After several hours and a bunch of erasing, Max declared that he was finished. They rolled the Lexan sheet, and the cylinder fit within the core frame.

Victory one.

Max and Chase picked up the helix itself from the floor to place it inside the outer cylinder. The helix was made from four spiral turns, each cut from one sheet of aluminum, and the four were riveted together to form the whole structure. Like a Slinky, it lay flat, but when lifted by its top spiral, it extended to more than four feet tall. Also like a Slinky, the helix tended to have a mind of its own. The two students tried feeding the helix into the cylinder from the top, Max holding it while Chase stretched his arm inside the cylinder and pulled the helix down into place. There was a lot of grimacing and sounds of metal stretching and compressing. If the two weren't careful, they'd bend or warp the helix beyond repair.

"Why don't you turn it on its side?" Stan said.

The two students looked at their mentor as though he had just invented a time machine. A second later they had the helix in place, and they lifted the structure upright.

"I just want to see if the balls roll down the ramp," Amir said, standing alongside a cluster of students jostling to see the robot core come together.

Amir placed a ball at the top of the helix and let go. It rolled all the way down the spiral to the floor.

Victory two.

Now it was time to fit the helix to the lines Max had drawn and then secure it to the Lexan cylinder. To do this, they had to drill through the thirty tabs spaced evenly along the sides of the helix and then rivet them to the Lexan.

When they started this next stage, it was already 8:30 P.M. Chase and Max held the helix in place. Colin and Luke kept the tools ready. Amir secured the spiral into position. Starting at the bottom, he lined up the spiral to the black marker line, hammered the first tab flush against the cylinder wall, drilled a hole through the Lexan, and riveted the tab into place. The ten minutes spent on this first of thirty tabs promised a very late night.

They fell into a rhythm: hammer, pound tab, pass drill, punch through Lexan and tab, pass rivet gun, *pop,* secured, rotate cylinder, next. Remarkably, they finished a few minutes past eleven o'clock. Led Zeppelin's "Stairway to Heaven" played on a pair of speakers as they riveted the last tab of the spiral ramp into place. Amir inserted a ball at the top, and it rolled down the helix as well as before.

Victory three.

Chase and Gabe started feeding balls into the helix to see how many it could fit. Twenty-four.

Victory four.

They then inserted the rolled aluminum inner cylinder down the middle of the spiral. The fit was snug, just as they wanted. Too much space, and there wouldn't be enough compression between it and the outer cylinder to move the balls up. Too little, and the balls would jam.

Now the final test: Would the balls ascend the spiral when they spun the inner cylinder? Eventually, a motor would drive this cylinder, but for now rotating it by hand should work. Amir gripped its top edges and turned. He strained to rotate the cylinder even a few inches. The balls wouldn't budge. There was too much compression, but even still, the balls should be moving at least a little. It wasn't working at all.

Amir looked at the students. Their eyes were wide, their mouths tight. They were silent. They were panicking. He was too, but he needed to calm them. "We might have to design an inner cylinder that's different."

The students remained silent.

Amir suggested that they could reduce the size of the cylinder. They could also wrap it with a tread to provide some grip on the moon rocks. Or maybe compression wasn't the way to go at all. A few teams had posted some comments on ChiefDelphi about using a spinning cylinder with brushes that batted balls up the spiral of their robot core.

This was an operating principle of *FIRST*: teams helping one another, whether by offering guidance on what they found was working or assisting other teams in their build, even though they would soon be competing against one another. "Gracious professionalism," Woodie Flowers called it. The principle was an expanded version of the old adage "It's not whether you win or lose. It's how you play the game."

"But we designed the whole mechanism for compression," Luke said, atypically downbeat.

The clock struck midnight, and they disbanded. After riding high for most of the week, they felt this defeat like a strike of doom.

Saturday morning failed to mark a turnaround. After a week where every single session had stretched past midnight, everybody came into the build room feeling groggy. When Chase arrived his eyes were red and his movements turtle slow. With his gray hoodie pulled over his head, he looked like he would fall asleep at his workstation at any moment. Amir was no better off. Sitting in a chair while ordering parts, he hunched forward, legs crossed tight. When he yawned, he turned away from his students, not wanting to reveal how weary he was. Gabe had a hangdog look as well.

Nothing was coming together. When they tried to inventory all the parts they had made, the server was down. On picking up the shooter parts that Valley Precision had machined for their practice robot, they discovered that several were missing, delaying a much-needed test. When they tried to mount the turret plate on top of the robot core, the metal warped. When Gabe thought he had the trigonometry right on the joystick controls, he found he was wrong.

Their mood worsened by the almost constant mentions from students to take a look at this robot or that robot that other teams had posted on ChiefDelphi and YouTube. Comments like "They have a system that's working, and we haven't even tested ours" prevailed in the build room.

Amir was punchy and short-tempered. When Stuart asked him where a particular electrical wire should go, Amir started chuckling. Everyone wondered what was funny. He explained, "The first thing that came into my head was, 'It's going to attach to your mom.'" Everyone started laughing.

"Well, Shaeer," Stuart said, unfazed. "Sounds like mid-thirties is the new high school."

Moments later, Amir directed one crew of students to get some tools and another to hit the machine shop, but they responded sluggishly. He shouted, "Hurry up, people. We're running out of time!"

When Stan arrived, they held a team powwow to discuss what to do with the robot core. Stan had left the previous night before they had discovered that balls didn't travel up the helix, and they now ran a demonstration for him. "It's tight," he said, shaking his head. "I think we're having height issues too. The whole thing's a lot more rigid than it needs to be. There's too much compression."

They decided their original design wasn't the way to go. Everyone liked the idea of batting the balls up the helix by attaching some kind of rigid brushes to a smaller, but still spinning, inner cylinder. "It could be a big push broom," Amir said. "They'll hit the bristles and grab and move up." Stan and some students left for Home Depot to see if they could find the right brush material to use. Chase searched the Internet for a tube to use for the new cylinder. They needed a solution, quick, or they would have a robot that could do little more than drive around the field.

As Saturday afternoon turned into evening and evening into late night, they continued working. Amir gave no rousing speech to motivate his students. He didn't need to remind them of the urgency. In ten days, they needed to ship. They still hadn't even finished designing their practice robot, let alone started assembling the one for competition.

There was no more resentment between those who stayed until the bitter end and those who didn't. Everyone was working hard, clocking a minimum of thirty to thirty-five hours a week in the build room. Some, such as Chase and Gabe, were putting in at least sixty hours, but like Amir had told them, they felt like they would get more from the experience because of their dedication. They were not quite yet a team, Amir thought that night, but in the collective struggle to finish in time, they were getting there.

Everybody knew there wouldn't be any epic breakthroughs that would solve all their problems with the robot. They just needed to work and work and work.

"I think we're going to get it done," Chase told Luke. "It's only how much is it going to hurt?"

Week Six

8 DAYS UNTIL ROBOT SHIP

I know. Velcro and zip ties are already the other women in
our relationship.

—EMILY WEST

Sunday was no longer a day of rest.

Problem creep was becoming pervasive on the practice robot. With each minor problem the students fixed, more presented themselves. With each solution, the SolidSeven amended their designs so that the parts machined for the competition robot would not repeat the problem.

The shooter was one of many elements that needed revision. That Sunday afternoon the team had received the missing parts to finish putting together the hood. They wired up the shooter. The programmers loaded the code to spin its motors. But when they fed balls into the shooter, they dribbled out, much as they had with their first wood prototype. There was not enough compression between the flywheel and the curved hood. Daniel and Yidi screwed layer after layer of rubber tread onto the inside of the hood until they found the right squeeze. In SolidWorks, Stuart decreased the size of the hood to achieve this same compression for the competition robot.

At 6 P.M., Amir called Emily.

"I know you're going to start working late," she said. "But that's not going to be tonight, right?"

"No, I'm coming home at eleven," Amir said, even though he had wasted much of the day fixing problems with the chassis design.

"Really?" Emily asked. She was fine with him working this Sunday. His days off during robotics season were not relaxing for him anyway. He stressed about what he needed to do with the robot and often ended up with a headache. However, she knew once he started working until the early morning, he would keep to the exhausting schedule until the robot shipped. This work ethic had been ingrained in him since he was a kid, whether pulling weeds on a landscaping job with his father or holding tools or a flashlight for his father while fixing the house. Sometimes Amir pushed himself too hard though, and Emily worried he was starting his sprint to the finish line too soon. "Today's not the time to start being really late."

He promised he wouldn't.

"We'll see," Emily said.

A few hours later, Amir called her again. They still needed to figure out how to get balls up the helix, and for every problem they fixed on the robot core, two problems popped up. He couldn't be home by eleven.

Emily heard the stress in his voice. He wasn't sleeping well, and with the pressure on him to see this team succeed, both for their sake and for their academy, he was not in a great place. "You've done this before," she said. "You're great at this. You'll do it again."

"I have a headache," Amir admitted.

"What's the most important thing that needs to happen right now?" Emily asked, trying to get him to see the big picture. Then she told him to come home when he was ready.

The team remained in the build room until 3 A.M. Before they left, a student wrote on the whiteboard in large block letters: "Team 1717 is epic." The statement was right next to the more mundane "Clean up room."

With eight days until robot ship, and everything they still had to do, the team continued to give potential donors presentations and tours of the academy. That Monday afternoon, Gabe was the tour guide for a UCSB professor, and by the end, the seventeen-year-old was openly yawning and stretching out his arms. When the professor left, Gabe said, "That was the longest hour and a half ever."

"Hour and a half?" Kevin said. "It was like thirty minutes."

Gabe retreated to his workstation. With the many open and closed loops in the turret manager, drive train, and other mechanisms, and all these loops running simultaneously through the controller, his code was crashing again and again. References being used didn't exist. Steps were missing in logic statements. Sometimes the code syntax was just plain wrong. Computers are literal, and if they don't have the right instructions, they don't know what to do and error messages sprout like weeds. To debug, Gabe ran the code in slow motion. It was like watching a car crash unfold, and he looked to see whether it was the dog crossing the street, the cell phone call, or the traffic light on the blink that had caused the sequence of events.

Amir and Stan sat beside the practice robot, trying to find a spot on the side of the robot core to place the battery. "Can't we put it on the trailer?" Stan asked. "We've got one, we may as well use it for something."

Amir laughed, a rare and welcome break from the stress. "That would be innovative." He moved the battery to the front corner, opposite the electrical board.

"Structurally, it's fine," Stan said. "Aesthetically—"

"It makes me want to vomit," Amir finished his sentence.

Much of the build room suffered from a similar standstill. The students had finished all the parts of the competition robot's wheel modules, but they couldn't put the modules together, or work on any other assemblies, because the team not only wanted their robot to operate at the highest level, it also had to be a mechanical piece of beauty. Even though they were far behind schedule, they didn't want to sacrifice the "fit and finish" of the robot. This was as important as any of the mechanisms they had listed in the Absolutely Need column on the first day of the season.

Amir wanted every part for the competition robot machined by Wednesday night. On Thursday they would send the parts to a company that anodized aluminum. The anodizing process would make the metal scratch-resistant and corrosion-free. Most important, the process, which involved dipping the metal into vats of sulphuric acid, allowed for the aluminum surface to be dyed gold and black, the team's colors.

Only on Friday, with four days remaining before ship, would they be able to start assembling their competition robot. Yet they still had not solved the problem of how the helix would deliver balls to the shooter, let alone built an operable shooter.

Later that night Stan oversaw Max and Stuart, alternating between them, as they revised the designs of the helix and shooter. He had CAD drawings in one hand and his eyeglasses in his other. Whenever he was anxious, which was pretty much all the time now, he placed one of his eyeglass stems in his mouth. "This is chaos," he muttered to nobody in particular.

Amir was growing frustrated as well. He was facedown on the floor, his hands deep in the recesses of the robot base, trying to figure out why some of the balls were sticking on the helix. He found that a part of the chassis was interfering with the balls and would need to be shorn off. Next he went in search of a little bolt they had machined. They needed it for the new inner cylinder, which they were hoping beyond hope would arrive tomorrow with a set of brushes to bat the balls up the helix. Unable to find the bolt, he stopped in the middle of the room, shook his head, and started to make his way to the machine shop to fabricate a new one. Before he disappeared, he turned back to Stan. "You were right, Stan. You win."

The students didn't know what Amir was talking about, but Stan did. The schedule he had shown Amir the day the plague hit had proven to be right. They were running out of time.

What was clear to the students, however, was how much effort and intensity Amir and Stan were investing in their team.

"It seems like they'll go to any lengths to see us succeed," Luke said to Chase close to midnight, with both their mentors still going strong. "That's what makes them good."

"My grandpa was an engineer," Chase said. "He told me that no missile ever worked before ten o'clock at night."

Luke and several others nodded as if they were well acquainted with such things.

"I heard a story once about the stealth bomber," Chase continued. "The engineers went somewhere to present it, and their bosses vetoed their design. They went back to their hotel that night, revised everything, and brought it back the next day, and it got okayed."

The team hoped what Chase was implying was right. Somehow everything would come together for them one of these late nights.

Tuesday, February 10, Amir spoke bluntly with his robotics team—or, more accurately, the handful of students who remained after the bell struck ending their 9 A.M. robotics class. "We're *screwed*," he said.

The statement struck Gabe and Chase and the others. Their teacher was always straightforward, and the later the nights went, the more pointed he would become. They were used to that. But then he usually couched his tough love in some humor. That morning, none followed, and he left the students crestfallen.

"We have to get our act together. You have to step up," Amir said. They needed to get the robot core to operate that day, and there could be no more major changes to any mechanism. If they had any significant delays from now on in, they were doomed. They still needed to test the competition robot before they shipped or they'd be stuck trying to fix mechanisms at their first competition.

When the team met in the build room a few hours later, Amir's candid talk to his students had the desired effect. The students bustled about, on task and focused. The mishmash of statements and to-do lists on the whiteboard had been erased. Now there was a single list of the different mechanisms and who needed to do what on which.

Before dinner the practice robot was lying on its side. Students were unpacking the cardboard boxes of newly arrived parts for the inner cylinder of the robot core. There was a clear polycarbonate tube, brackets to mount it to the chassis, and long thin strips of aluminum with 6-inch black-plastic bristles sticking out from them.

"We need to get this running tonight," Amir said, holding the motor that would spin this new cylinder.

Almost as if on cue, Emily arrived at the build room to provide moral support. Now that her morning sickness had passed, she was there more and more. She knew how essential it was that they get balls to travel up the helix as soon as possible.

The team spent a couple of hours constructing the new inner cylinder. They screwed the two lines of brushes onto the tube so that they were opposite each other and ran the tube's length. Then they mounted the cylinder inside the spirals of the helix and connected a motor to it.

By 8 P.M., Stan wanted to leave, hoping to see his young daughter before she went to bed for a change.

"Stan," Amir said. "I'll do whatever you want as long as you stay until this thing moves."

Finally, they tilted the robot upright. The students sat in chairs or stood in a semicircle around the robot to watch the test.

"Turn it on," Amir said.

Kevin sat with the operator console in his lap. He moved a joystick, and the inner cylinder started to spin. "You want max speed or what?"

"Go for it," Amir said.

The tube with its two long strips of black brushes started to spin so rapidly that it looked covered with brushes. Amir then placed a ball into the robot base where the helix started. The brushes took hold of the ball and thwacked it up the spiral ramp in less than four seconds. A round of applause followed.

"Okay, faster, faster," Gabe said.

Amir fed more balls into the robot base. They too traveled up the spiral.

Then little bits of plastic started to scatter about the floor. "Wait," Emily said. "The brushes are shedding."

Kevin cut power. Scores of black bristles, looking like bobby pins, littered the inside of the robot.

"Why's it shedding?" Amir asked.

"It's coming out where we cut it," someone said.

"Fire up the hot-glue gun," Amir ordered.

"Hot glue!" several students said.

"You can never get too much hot glue," Chase said.

The team was encouraged. The anxiety over the core had lessened a little, and the hijinks began. Students began taking the bristles, bunching several together, and massaging one another's backs and scalps.

"This is good. It's like a drug. It's, like, hair-ajuana," Colin said in his best surfer-dude accent.

"It's like comb-caine," Stuart added.

"It's like hair-oine," Colin said.

"How about meth-stache-amphetamine?"

"Man, that's a reach."

Once they cleared the loose bristles out of the robot and hot-glued the strips of brushes, they were ready for test two. The inner cylinder started spinning. Amir fed it the first ball. It ascended the spiral with the thumping sound of a washing machine.

"Okay, more balls. Let's do it!" Amir said. "Now!"

Gabe fed more balls into the robot base, and they rose up the helix. No bristles flew out. Five balls. Ten balls. Fifteen balls. It was working.

"Are we full yet?" Amir asked.

"No!" the team yelled.

More balls filled up the helix. Twenty balls. Twenty-four balls.

With the turret plate and shooter not in place, Amir pulled a ball out from the top and another pushed to the top of the helix. Students laughed, clapped, and smiled widely. Stan beamed.

"Keep hope alive!" Amir said.

"Yes. We. Can!" several students called back in unison, echoing the new president's campaign slogan.

Amir sighed and sat down next to Emily. "Thank God it delivers these balls."

The team fixed the intake roller to the front of the chassis and wired the motor. This was the first time they were running the whole ball-collection system on the practice robot. When Amir rolled a ball toward the front of the robot, it became jammed within the base. He got down on his stomach again and found the source of the problem: a thin plate at the bottom of the chassis.

"I hate that plate," Amir said.

"I hate it too," Stan said. "I think we should try removing it."

Andrew hurried to the machine shop to retrieve a hacksaw. Gabe and Kevin tilted the robot on its side, and a few moments later they had removed the plate.

"I'm leaving now," Stan said.

"Are you more, less, or neutrally encouraged?" Amir asked. "I'm *way* more encouraged."

"I'm reasonably encouraged. I think tonight you should be thinking about the feeder," Stan said, referring to the wheel that kicked balls from the top of the helix into the shooter.

"Thank you, Stan, for staying."

Just before they started test three, Amir reached inside the robot base to inspect something. He warned his students not to turn anything on, but Gabe, who was at the joystick control, didn't hear him and was behind the robot, unable to see Amir. He turned on the motor that spun the inner cylinder.

Amir yanked back his arm and said in a tone that was more intense and serious than any all season, "DO NOT turn it on, I said."

"Sorry," Gabe said, shoulders sagging.

"He didn't hear you," Emily said. "Gabe, we still love you."

"I don't love you," Amir said, smiling. "I almost lost my arm."

They started again. The intake roller and inner cylinder both spun quickly. Students rolled balls toward the robot. The spinning intake roller sucked them inside immediately, and the balls started moving up the helix ramp.

"Team 1717 is back in business!" Amir said.

———

The next afternoon, after a long frantic session of machining by Stuart and Stan, the team circled around their practice robot again, the shooter assembled and in position. Six days until ship.

Several students lifted the robot onto a sheet of regolith and positioned their mock trailer 15 feet away. The adjustable hood needed to be retracted by hand, and the lazy Susan was not yet powered by a motor, but otherwise the shooter was primed and ready.

Emily was present again, dressed in a black trench coat and black beret. The team spent so much time in the build room that it was easy not to notice that the weather had turned cold. It was 45 degrees Fahrenheit outside, frigid for Goleta. Amir was still in shorts and a light-green sweatshirt.

He bent down below the hood. Through a hole a student had cut into the outer Lexan cylinder of the robot core, he inserted the kicker wheel: a 4-inch-wide PVC pipe wrapped with tread and attached to a small motor. This kicker wheel would propel balls from the top of the helix into the shooter. They had yet to figure out a place to mount it permanently, so for these tests, he would just hold it in place himself.

"Shoot just one," he instructed Kevin, who was at the joysticks.

There was a distinct *swoosh* as a ball hit the kicker wheel, spun upward to the flywheel, and was propelled out of the hood into the trailer. "Another one," Amir said, rotating his head back and forth between the robot and trailer like he was at a tennis match. Another *swoosh,* another score.

"Go for it," he said.

Swoosh.

Swoosh.

Swoosh.

A continuous stream of balls launched from the hood, arced through the air, and dropped into the trailer basket. Of the twenty-four in the helix, four missed their target.

Amir walked over and hugged Emily, then stood by her side, his hands jammed into his pockets. He was all smiles. "That was pretty freakin' sweet! Now I just hope there'll be balls available on the field."

The team reloaded and ran the test again. Balls arced through the air in a blur of orange and purple and green. "Four misses from that distance?" Amir said. "That's ridiculous. I love this thing. I would…"

He stopped and turned to Emily. She looked at him, eyebrows raised, egging him on to say it. "I was going to say I would marry it, but you were standing right here."

Emily laughed. "I know. Velcro and zip ties are already the other women in our relationship."

The team continued testing, adjusting the hood angle and flywheel speed. Near the end of the night, Amir asked to see "the death shot."

Chase and Luke pulled the trailer flush up against the front of the robot. The hood was angled down for up-close shots. Another student held on to the kicker wheel.

"This will see how we do as a power dumper," Amir said.

"Flywheel on," Gabe called out.

"Full speed," Amir ordered. "Fire!"

Twenty-four balls bombarded the basket. High fives and smiles filled the build room. As the team filed out before midnight, every major mechanism on the practice robot was now operable. Finally.

The next afternoon at Valley Precision, a nondescript, single-story building in a row of the same outside Goleta, Amir and five students watched some of the final sections of their competition robot being created. The team had fabricated most of the small or intricate parts of the robot. But the "big stuff"—as Amir called the chassis, the helix, and the shooter hood—were being cut and bent inside this shop. The Valley Precision machinists entered the final designs the SolidSeven had created in the 3-D drafting software. Then they laid down a sheet of aluminum onto the bed of the CNC punch-press machine and pressed a few buttons. Off the machine went, making a series of exact and complicated punch cuts into the aluminum within a few minutes.

The team spent most of their six hours at Valley Precision that Thursday doing grunt work in one of the few open areas in the shop. Crowded together, sopped in sweat, they grinded down, sanded, and filed away the sharp, burred edges of each part after it came out of the punch press, getting them ready for a pickup from the company that was anodizing them. Amir expected the parts back by Friday so they could start to assemble the competition robot.

The team left Valley Precision happy with what they had accomplished. "We greased the wheels of productivity," Amir said.

After dinner on Thursday, they filed back into the build room. The team had cleaned the entire space, organizing parts and tools in bins or on the shelves in marked Ziploc bags. They had even vacuumed the floor.

"This is awesome," Amir said. "Now let's talk."

Everyone grabbed a chair, and they all circled around Amir, who sat in front of the open door. He stretched out his legs, locked his fingers behind his head, and exhaled loudly. With the long Presidents' Day weekend, the team had Friday to Monday to work full-time in the build room. Tuesday they would ship. "I want to discuss what the next four days are going to look like so that we can both have some calm but also a little bit of the necessary anxiety we'll need to finish."

He turned to the electrical team, comprising Angie Dai, Isabelle D'Arcy, and a few other students. Without their work wiring all the motors and sensors, Amir said, the robot was just a bunch of useless parts. "I want you to have an attention to detail like you've never had before. We're going to have so many wires in this robot that I don't want to have to deal with wires not being soldered on correctly or put in backward. We're getting the competition robot tomorrow, so everything you crimp has to be done the right way. Last year we had three major electrical failures that made our robot sit still during an entire match. I'm issuing a whole different level of quality control this year because the stuff we're building now is so crazy. All those little wheel drives, they have to be right within a couple thousandths, or else the gears fail."

With Angie and the others adequately frightened, Amir moved on and recruited two volunteers to drive down to Los Angeles to pick up some additional hardware they needed. Then he turned to Chase and the drive-train crew. A conversation over the color of the shooter flywheel and how its red needed to "pop" interrupted his flow. "If we're this efficient," Amir said, "we're never going to finish. The drive train will be here tomorrow, and my intention is to start from the ground up."

The room went silent. He detailed each and every aspect of what they needed to do, as well as the tasks of the robot core and shooter crews. He pointed to the programmers. "I want you guys to start coding up some stuff." He talked shipping forms and rules for what they could bring to competition. He told them that they needed to get their homework done and alert their teachers that they might miss class Tuesday, the day of ship. He warned them not to drive at night if they were too tired and asked them to take care of one another. It was a massive download of information, guidance, and instructions, as if he was clearing his mind of everything that had been churning away inside of it.

It was time for the final push, and they were as ready as they would ever be.

96 Hours Until Ship

Can they just give us the trophy now?

—ANTHONY TURK

It was Friday the thirteenth. No two parts of the competition robot had been assembled, and there were more than eighteen hundred total.

The day started triumphantly. Kevin and Luke returned from Oxnard, entered the cafeteria where everyone was gathered, and held above their heads the black, anodized chassis. It was a beast of metal, and the team cheered. Once back in the build room, they filled a whole corner with cardboard boxes containing anodized robot parts. With most black like the chassis, the gold shooter hood and its red flywheel stood out.

"We've got a lot to do and not much time to do it," Amir said before everyone hunkered down to work. He wanted to focus on the competition robot, but there were some design issues to be resolved on the practice robot first.

Throughout the day, Amir was in bounce mode, moving back and forth across the build room. One moment he knelt down by the practice robot with Stuart, searching for the right place to mount the shooter's feeder wheel. A minute later he was advising the electrical team on what to do with the twisting snake of sensor and motor wires on top of the turret. They needed to find a way to keep these wires from tangling when the turret rotated. Then he was off to speak with Gabe about the camera placement for their automatic tracking system. "This is mission-critical," Amir said.

Everything was mission-critical now though. Assembling the wheel modules for the competition robot should have been easy, but they discovered that the anodization had added a layer, less than one-sixth the thickness of a human hair, to the parts. They had designed everything with such precision that now the parts didn't fit together. Chase and the drive-train crew needed to sand them down to fix the problem. More time delays.

"We're not leaving tonight until we get these done. I know it's frustrating," Amir said.

Then he was off to watch another test run of the shooter. They were timing how fast they could empty the helix of a full load of twenty-four balls. At first they were at 8.4 seconds. Then they switched the transmission on the kicker motor to add more speed and less torque. Afterward, the balls emptied in 6.2 seconds.

"With this ability to shoot and this much ball storage, we don't even need to bother shooting at a distance," Amir said. "We just wait, wait. Get right next to somebody, frickin' shoot it, and move on."

Amir and the students increasingly talked about game strategy and their upcoming regionals. It was becoming real to them that they would soon see how their PenguinBot stacked up against the competition.

Talk of the regionals seemed to accelerate Amir and his students. By midnight though, some were napping underneath their workstations, jackets used as makeshift blankets. Others, still in their chairs, rested their heads on one another's shoulders, fast asleep. After finishing the wheel modules, Chase lay down on the regolith. He was so exhausted that he fell asleep with his eyes slightly open.

At 1 A.M., the team stumbled out of the build room to their cars and eventually to their beds. Gabe found little rest when he returned home. The hot-water heater had burst and flooded his house. His mom and grandmother were struggling to get everything onto higher ground while making frantic calls to insurance agents. Barely able to stand, he pitched in to help. His father's absence was felt acutely. When Gabe did manage to sleep, he had a bizarre dream that it was his eighteenth birthday, and he went with his theater crew to a strip club in Santa Barbara. When he awakened in the morning, he didn't know what to make of it.

At least the dream wasn't a repeat of the one he'd had recently, where he walked into the build room to find Stan alone and huddled over in a chair, stressed out and weeping into his hands.

On Saturday morning Chase dragged himself out of bed. He was shivering from a fever but was soon out the door. His mother watched him leave, amazed that her son, who had always needed picking up after and prompting to get his homework done, was now self-sufficient. He woke up on his own, juggled robotics and school, and never uttered a word in complaint, even sick as he was.

Before heading to Dos Pueblos, Chase stopped at the flower shop. It was Valentine's Day. He doubted he'd have time to see his girlfriend that night, and he needed to make a preemptive strike. He picked out some flowers and chocolates to be delivered to her house. The sales clerk suggested he add some balloons that read BE MY VALENTINE, but Chase shook his head. "Please, miss. I'm already on the USS *Mega-Cliché*. I don't need to throw myself overboard."

Chase was not the only one on the team trying to keep his life outside robotics from falling apart. A few minutes after he arrived at Dos Pueblos, the sounds of a string instrument echoed throughout the build room. A few students followed the music to the nearby wood shop, where they found Andrew taking a break from the lathe. Sitting in a chair, a cello upright between his legs, a bow held in his right hand, he played Antonín Dvořák's cello concerto. He had sheet music perched on a table to his side and seemed oblivious to his surroundings—or his expanding audience—as he played the piece flawlessly. Eventually Stan came in, needing to use the drill press. Still, Andrew played. When he finished the piece, he said, "This is the one time I've been able to practice in the last month. I'm a soloist with the Santa Barbara Youth Symphony. I need to play." Then he started from the beginning.

Eventually, Amir ended the concert by calling Andrew back to the machine shop to refine some of the shooter parts. By the looks of it, little progress had been made the day before. Most of the cardboard boxes of anodized material remained untouched. Amir kept walking around the build room, chanting to himself, "Finish...Finish...Finish."

One of the electrical mentors crossed his path. Danny Lang, a Raytheon engineer, was a bear of a man, twice as wide as Amir. Out of high school, he had been recruited by several Division I colleges to play defensive end, but he declined the offers and studied engineering instead. He'd been drawn to be a mentor for the team because of a shop teacher who had once inspired him.

"Stressing out?" Danny asked Amir.

"Yes," he said, worrying over a delay in the arrival of the pair of students who had driven down to Los Angeles to pick up some parts. Every hour they weren't back was an hour less to assemble the competition robot.

Suddenly, Gabe pushed back his chair, stood up from his workstation, and said, "I give up. I quit." He stormed outside.

As always, the programmers were juggling several sections of the code: autonomous, turret, and now traction control for the drive train. The software itself was mind-bogglingly complex, and in their rush to code, they ended up with bugs everywhere.

Gabe was also aware that with all the developing mechanical problems, he would have a short window of time with the competition robot. This meant that their tests would be last-minute and, given everything that still needed to be done, he would still be putting a lot of hours in after ship date, particularly on refining the shooter. There was no limit on the time programmers could work after the build season. As long as the practice and competition robot were mechanical twins, they could continue testing their code on the former and expect it to work the same on the latter. This additional work did not excite Gabe. He missed the theater and some semblance of a normal life. He missed sleep most of all.

For the moment, he put that out of his mind and returned to his workstation. To blow off steam, he wrote a quick program, his fingers striking hard on the keyboard. He then called Amir over and pressed a button. A single message in bold black-and-red letters flashed repeatedly on the screen: "Stick It to the Man!"

"I wrote a program to describe the entire day," Gabe said.

Amir smiled and continued onward. More problems presented themselves, but through sheer effort and time, the team advanced the assembly of the competition robot.

Late that night, the programmers solved the problem of what to do with the 15-second autonomous period of matches, in which their robot would have to operate on its own, without pilot control. Gabe had written code that made the controller act like a recorder. Say they wanted to drive forward and then spin around to avoid shots from human players during autonomous. On a practice field before competition, Chase would pilot the robot to do that while the robot's controller recorded the joystick inputs from his operator console. During the autonomous period, they would have the controller replay those moves. It was beautiful in its simplicity and versatility.

Amir and a handful of others didn't leave the build room until 2:30 A.M. The next day promised to go even later, and it was certain now that on Monday night, their final night before ship, the team would have to go without sleep if they were to have a chance of finishing the competition robot before the FedEx truck arrived.

Slowly, visitors filled up the build room. Some came alone, others in pairs. They were in their fifties or sixties, mostly women.

They didn't know what to make of what they were seeing early that Sunday afternoon. A low black frame with wheels stood on blocks. Kids with bleary eyes huddled over computers and sat cross-legged on the floor assembling parts. They had left power tools scattered on plywood workstations and amassed a bunch of multicolored balls in a hexagonal basket made of wood and PVC pipes. From a closet in the back of the room came the most dreadful, persistent whine.

The students didn't know what to make of their visitors either. Surely, there was a better time for a tour of the academy than in their make-or-break moment. The team tried to ignore their presence. Chase sat on the edge of his workstation, debating with his teammates how they should put on the bumpers. Amir disassembled a motor transmission, his knuckles bloody from the effort. He avoided any interaction, trying to stay on task, but that became impossible when Parm Williams, the representative of the Women's Fund, told Amir that her members were ready.

"The grand presentation," Amir said, putting his motor down on a workbench. He gestured for his students to gather.

The team ran through a demonstration of their practice robot. "This is the prototype," Amir said. "The real one is being built up now. Hopefully, in a day and a half, it'll exist."

Next to Amir, Parm Williams held out a bank check. She started, "I think this academy is so important to the future of education. We'd like to thank you for doing what you do." She passed the check halfway to Amir. "I just hope this will help you raise the matching funds you need by the November deadline. On behalf of the Women's Fund, we'd like to present you a check for $150,000."

"Thank you so much," Amir said, his words lost in all the clapping. He then crossed the room to a tall shelf by the door. Stretching up high, he grabbed some blueprints of his future academy. He started showing off different aspects of it to his visitors. His excitement was unmistakable, and his students, who knew and respected his vision, were once again reminded that this season was not only about them. They were working for something greater than themselves.

"Okay," Amir announced to his visitors. "I'm off to build robots."

Through the rest of the afternoon the room buzzed with activity. After dinner, a burst of raised voices came from the corner where the electrical team worked. Mentor Danny Lang had invited another engineer to help the team over the weekend, and he was supposed to be showing Angie, Isabelle, and several other students how to wire a section of the electrical board. Instead, he was doing all the work himself.

"We should be allowed to do it our way," Isabelle said, her feet wide on the floor like she feared she might fall down. At the beginning of the season, she had been reluctant to even speak at presentations.

"This is complicated," the gray-haired engineer said. "It's easier if you watch me and learn from me. If you do it, we'll be here all night."

Isabelle glared at him. Tears formed in her eyes, and her cheeks reddened. This was not how she was going to learn; this was not how she had come to expect to be treated. Learn by doing. That was the academy. She knocked back her chair, grabbed her backpack, and rushed out of the room.

The rest of the electrical crew refused to back down. A minute later, the engineer threw his screwdriver into the air and said, "Fine. You take care of it. It's your project."

Amir called Danny over. "Look, this guy is being kind of a jerk. He needs to know we're here to educate students."

The engineer left and never returned.

As this storm blew over, Max and Turk put together the structural frame that would house the helix. Stan was helping them. "Can you believe in about forty hours this is going to be ready? With its little omni-wheels driving into the crate perfectly."

None of them were sure that would be the case, but it was nice to daydream about before the sheer volume of work ahead overwhelmed them. Their competition robot was still in scattered sections throughout the build room.

Over the rest of the night, Amir zigzagged among crews. They installed the intake roller on the chassis. They secured the shooter to the turret plate. They finished the mounts and transmission for the kicker wheel. They fixed the wheel modules onto their base plates. They assembled the separate spirals into the full helix.

While most of the team worked on the competition robot, the programmers stole any time they could get with the practice robot. They took data on the best flywheel speeds and hood angles for making shots into trailers at different distances.

At 3 A.M. the students were stumbling around in a daze.

"I say clean up and get organized," Amir suggested a half hour later. "If we can't work efficiently tomorrow, we're toast."

As they staggered off to put away tools and clear their work-stations, the whole build room began to shake. Then there was a crash of metal outside.

They opened the door to howling winds. They hadn't even noticed the approach of the storm. Turk walked outside, leaning forward so as not to be blown back. "The wind out here's crazy," he said.

While the students packed up to leave, Amir was thinking about what he hoped they were getting out of this season. He wanted them to learn about engineering and see how exciting it was to design and build a machine with their own two hands. But more than that, he was coming over to the idea that the real purpose of this experience was to show them the kind of focus and hard struggle it took to create something extraordinary. Teaching from a book or for a test would never show them how to obtain the best out of themselves.

Nights like this one would.

Twenty-four hours until ship. Everybody arrived back in the build room by lunch. Some walked stiffly as zombies, vacant stares on their faces. Many had parents in tow. They came inside to see the robot—or rather, its separate parts about the floor. They also unloaded supplies from their cars to see their kids through their longest day: platters of cupcakes, Rice Krispies Treats, fruit, bags of chips, pretzels, Doritos, and M&M's. Cases of bottled water and soda. Most important, they brought AriZona iced teas—or "Zonas," as the team referred to them now. The drink was in such demand that it had become its own currency: One Zona equaled one dollar.

As the parents filed out, students herded around a computer to watch a practice match between teams in Los Angeles that had been posted online. Balls were flying everywhere, and the robots seemed to be lost on the white regolith field. "This game looks so chaotic," Chase said.

Amir glanced over his shoulder. "I'm hoping all of our hard work will make it less chaotic."

With that, the final build session began. Amir made a circuit around the room, moving from shooter to frame, bumper, helix, electrical, and anything else that needed his attention. When the lead screw on the shooter hood made a grinding noise as it moved back and forth, he told a student to shave down the shaft. When the new core measured a little shorter than they wanted, he decided spacers would do the trick. When they were marking where the tabs that secured the helix to the outer cylinder went, Amir draped the cylinder over his head like a dress and started making marks with an X-Acto knife. Even then, with half his body enveloped in robot, kids were asking for help, needing to know the diameter of a part they were cutting on the lathe and seeing if he could come check on the turret plate.

For every problem Amir managed, his students, working together, solved many others themselves. After six weeks of the build season, they now had the experience and confidence to work on their own. Chase was dividing his time between the machine shop and the drive train. Gabe was spending most of his time programming, but with his theater-production experience, he was also proving helpful with mechanical issues. Stan and the other mentors pitched in when needed.

When Amir stopped long enough to survey everything going on without him, he was encouraged. There was no sense anymore of different work crews separated out in the build room. His students all mixed easily now, helping one another out. One student passed another a tool without even the need for explanation. Everyone looked focused and determined. It was all beautiful to see.

After dinner, Amir gave the team a short pep talk. "All of our best-laid plans have kind of fallen apart to some degree, but we're going to finish. Now back to work."

Emily had come by to eat with the team and was now leaving. "Are you okay?" she asked.

"You're coming back, right?" Amir said, almost desperately.

"Yeah, I'll come back."

"Then I'm okay."

Soon after she left, Amir announced, "We're doing the helix." Three hours of painstaking, meticulous work followed to make sure they did not scratch the crystal-clear outer cylinder while installing the helix. Even at this late hour, they were concerned about aesthetics.

As midnight approached, Amir said to his team, "I hope you're enjoying this process. You're going to see the robot come to fruition."

On they worked. Now and again, the students took a break. At one point, they brought out the air hose in the machine shop and held a Sharpie pen in front of it, creating a makeshift spray-paint gun to tattoo their stomachs with bull's-eyes. More often, they just made fun of one another.

"After the success of last year's team," Chase said, standing by the lathe, "I'm feeling the pressure."

"Personally, I don't believe in pressure," Turk joked, leaning against a chair, his chest puffed up.

"Yeah, we'll see what happens when you get a moon rock in your hands," Colin said.

"Dude, have you ever seen this guy take a test?" Chase asked.

"Hey, come on, now," Turk said, smiling.

"Turk covers the paper in sweat," Chase said. "He holds rags so the pencil doesn't slip out of his hands."

They spoke of the competition and their expectations for the season. Luke predicted, "We'll win in L.A. and go to Atlanta."

"There's no margin for error," Turk said.

"If we win in Atlanta," Chase said, "I'll get 'Team 1717' tattooed somewhere on my body."

"How about on your knuckles," Gabe suggested.

Overhearing the conversation, Amir declared, "Chase, I forbid you to get a tattoo."

Midnight passed. Then one o'clock. Then two. Then three. The robot was coming together. The drive train was operable. The ball collector and helix were set in place. Next they had to work on the inner cylinder for the robot core and the turreted shooter. But everybody was dragging. They had depleted their food supplies, so some students drove to Freebirds. They spent more than a hundred dollars on burritos, nachos, and quesadillas, to which Amir said, "You realize what that must be like as it tries to digest in your system at this hour." He had forgotten the iron-clad stomachs of high school students.

Outside a steady rain started to fall, but the build-room door was closed. They were sealed off from the world now. By 4 A.M., the room grew still and empty. Amir was fiddling with a transmission on a workstation, spending a half hour doing what should have taken a few moments. Now and again, he had to shake himself awake. Turk was painting "1717" on the bumpers. Gabe was hunched over his keyboard, a Zona by his side. Most of the rest of the team was in the room next door, either crashed on the floor with comforters and pillows or taking a momentary retreat from the robot to play a video game.

After a brief rest at home, Stan arrived back in the build room at four-thirty. Those awake applauded his return. The electrical team rallied and went through all of their connections. Gabe called Chase over. The programmers were feeling comfortable enough with their turret manager to give their code a full test on the practice robot, and they needed a pilot. With bloodshot eyes and hands shaking, Chase took the controls and directed the practice robot to the side of the trailer. He started to rotate the turret with the joystick, but Gabe told him to stop. "Wait. Let it find it."

Chase released the joystick. The turret locked onto the trailer and started shifting left until it was centered for the shot.

"Fire," Gabe said.

Chase hit the joystick's "fire" button. Balls streamed into the trailer. No misses.

"Can they just give us the trophy now?" Turk asked.

Two nearby students, who were sleeping at their workstations, didn't even stir on their pillows.

Next Chase tested the adjustable hood. The turret once again locked onto the vision target on the trailer. When Turk pulled the trailer farther away from the robot, the hood automatically shifted back, providing more arc when the balls propelled out.

"It's doing a lot of things mediocre," Kevin said, eyes glued to his computer screen full of code.

"We need to get it to do everything well," Gabe said.

A half hour later, Amir was recharged. He had taken a short walk and returned to the build room, asking, "Who do I need to sign out of first-period class?"

Hands shot up into the air like salutes.

"I'll sign you all out."

Afterward, Gabe collapsed. He crawled underneath his workstation, gathered up his jacket, and fell asleep, his feet sticking out from underneath. Chase lay down next to their unfinished competition robot as if to guard it. His head rested on a half-filled water bottle. He was so tired that he fell asleep with his eyes open again. Yidi thought Chase was still awake and for a second couldn't understand why he wasn't responding to her when she asked a question.

By 8:30 A.M. on Tuesday, February 17, everybody was awake. Chase stumbled to his feet and, feeling cold, jammed his hands into his front pockets and nearly pulled down his pants. "Easy there, guy," Stuart said. Occasionally, the door swung open, revealing the harsh glare of the sun. They had survived the all-nighter.

Amir stood before the competition robot. He was so tired he was swaying like a drunkard. In his mind he ran through everything they still needed to do.

In a frantic three hours, the robot seemed to come together like someone had waved a magic wand. The inner cylinder, now with four lines of bristles instead of two, was set inside the helix. At first it was wobbly, but after it was given a quick adjustment, balls shot up the helix smoothly and quietly. Then the kicker wheel, turret plate, and finally the shooter were mounted. Other than instructions from Amir and requests for this tool or that tool, they seldom conversed. Once they had finished, the electrical team swooped in and wired all the sensors and motors driving the turret, hood, kicker wheel, and flywheel. The programmers then loaded their code. Everything spun when it should spin, and with its black anodized body and golden hood, the robot looked like a work of art.

The team ran through a systemwide check at 11:30 A.M. Gabe powered down the robot, then restarted it. Their work of art wouldn't budge, not an inch. Gabe used a multimeter to measure if electricity was flowing. He touched almost every mechanism on the robot, but that wasn't the problem. Amir slumped down into a chair in front of the robot. They expected the FedEx truck to pick up their robot between noon and 3 P.M. They were out of time.

Everyone gathered around the robot.

"Why isn't it working?" Amir asked, pulling his baseball cap down in front of his eyes. He was too tired to troubleshoot.

"Nothing's working," Gabe said unhelpfully.

Shoulders sagged. "Oh no," several students said at once.

Gabe rechecked everything, then he bent down by the drive train. He stopped. The build room was hushed. He turned and explained that a wheel sensor was dead. He checked the one on the opposite side. It was dead too. Problem isolated, but this wasn't an easy fix. They had to remove pulleys and belts and reinstall and calibrate the sensors. Forty-five minutes later, they had the robot operating again.

"Seventeen-seventeen! Seventeen-seventeen!" the team chanted.

It was past noon. The students kept a careful watch at the front of the school for the FedEx truck. Once it arrived, they needed to crate up the competition robot, finished or not. During lunchtime, the whole of the Dos Pueblos High School converged on the build room. The principal, teachers, athletic director, bands of students—they had heard of the all-nighter and this amazing robot. They wanted to see it for themselves. The machinists from Valley Precision also arrived, as well as a host of parents. Amir offered demonstrations, even though his programmers were desperate to have more time with their robot. Amir wanted people at his school to pay attention. He wanted students to think that what the robotics team was doing was cool.

With Kevin at the joystick controls, Amir demonstrated the "commander move" and how the robot could maneuver in any direction at any time. The robot swerved about the build room and gathered balls. They spiraled up the helix with the *swoosh-swoosh-swoosh* of the spinning inner cylinder.

"Let's do the über-dump," Amir said.

The robot pushed up against the trailer and unloaded almost twenty-four balls in the span of a few seconds.

Between these demonstrations, they continued to test the robot. At two o'clock, Amir got the call. "FedEx is here!"

They still had motors to test and mechanisms to fine-tune, but they had no choice. The team rushed to get the robot into its black crate slapped with shipping labels. The FedEx driver lowered the loading platform, a skeptical look on his face as the team ran around frantically. Chase and Colin strapped the robot down, making sure everything was tight. The team crowded around, snapping photographs, before they screwed the crate lid shut.

The next time they would see their robot was in Long Beach, California, for their first regional competition in early March. The team watched the FedEx truck roll away. They didn't return to the build room until the truck disappeared around the corner of the parking lot. Amir found the moment bittersweet. He imagined this was how he would feel when he sent his daughter off to college, proud of who she was and hoping she was prepared for what she would soon face.

Forty-six days had passed since the kickoff. The Dos Pueblos team had not compromised in building the robot they thought would best meet the challenges of the game. "Go big or go home," Amir often said. No question they had gone big. Now their PenguinBot and their team would be tested in competition.

Los Angeles Regional Competition

LONG BEACH ARENA, MARCH 12-14
PART I

We want to be on a way higher plane.

—AMIR ABO-SHAEER

On Wednesday afternoon, March 11, a tour bus rolled to a stop in the narrow parking lot in front of the build room. Students came out to see their ride to their first regional in Long Beach. Most of them wore their sleek black Team 1717 jackets. As usual in Goleta, it was a mild 55 degrees Fahrenheit with low humidity and a gentle breeze. The jackets were unnecessary, but they were new and the kids felt cool in them. In the slit pockets on the arms some stored their cell phones, others tools, still others bags of candy given to them by their high school's cheerleading squad. Each bag came with the same note that used the sweets' names to craft words of encouragement: "Although some may think you all are a bunch of NERDS, we know there's nothing wrong with being SMARTIES. But, whatever you do, just don't be an AIRHEAD! Good luck in Long Beach!"

By 6 P.M., the students had clogged the cargo compartment underneath the bus with backpacks, duffel bags, and plastic bins filled with tools and spare parts. Amir made one more sweep of the vacant build room. On one wall hung framed photographs of the D'Penguineers from previous years' competitions. The absence of any faces from this year's team reminded Amir that everybody about to get on the bus outside was a robotics rookie.

He locked up. When he turned toward the bus, he found Gabe wearing two mangled moon rocks formed into a helmet, chin strap included. Amir broke down laughing. "If you have that on in the pit, you're getting murdered," he said, then turned to the team. "Please save him from himself."

"He wanted to wear it at school while riding around on a scooter," Yidi said, laughing along with the rest of the team.

Amir turned to Gabe. "Is that good for my engineering academy and its cool factor?"

The hat came off, and they boarded the bus. The students called for Amir, "Speech! Speech! Speech!"

Their teacher put on a sober face, like he was preparing to make a profound statement. He looked down to his side, where someone had left a half-empty bottle of water on top of a box of spare electrical parts. "We're not leaving until the person who drank this water comes up and claims it."

"Turk!" several students yelled, naming their usual scapegoat.

"I hate all of you," Turk returned.

After a short, unsuccessful interrogation of the team to find the real culprit, Amir told the bus driver, Roy, that they were ready. Roy closed the door and drove out of the parking lot.

"Let's go, baby!" the kids yelled.

A half hour into the ride down Highway 101, the sun settling over the Pacific, the kids settled down. Amir sat in the front, next to his wife, his black Team 1717 baseball cap pulled tight over his head. He had already given a big speech to the students earlier that day in class, saying, "A lot of the build season, you spent with the people in your group, working on different parts of the robot. We're not doing that anymore. Now we have to be a team. Everybody has to be there and everybody has to help out. Sometimes it's going to be a boring job, and sometimes fun, but you're going to have to do it and realize it's for the better of the team. If we do that, we'll have a better robot and a better experience in the end."

With his students broken into their usual groups on the bus, Amir thought how hard it was going to be for them to come together. Most of them had never participated in a team sport, didn't know how important it was to work as a single unit. Amir had been much the same in high school. Being a drummer in the marching band doesn't exactly prepare you to be a coach, he knew.

Amir recognized how burned out they all were from the many long days and nights and wondered if he had pushed them too hard. Even after they shipped their competition robot, the team had continued to spend many hours together working. Every day Chase and Kevin practiced driving in the cafeteria. The team machined a bunch of spare parts, prepared promotional materials on their PenguinBot, and readied their flight suits with their names, team logo, and sponsor patches. They also ran their practice robot through a series of tests, which prompted them to alter a few mechanisms to improve its performance, changes they would also make on the competition robot.

Everybody had been busy, but as usual, with the exception of Amir himself, nobody had put in more hours in the past few weeks than Gabe. From scratch, he wrote the traction-control code for the drive train. He spent days improving the tracking software for the turreted shooter. He cleaned up the code he had written during the rushed build season so that it would be easy to debug once they were at competition. He kept staying later and later in the build room and then working later still in his bedroom. At the end of February, while spending another long night working at home, he broke down in tears from the stress and persistent lack of sleep. On learning of this, Amir addressed the team, saying nothing of Gabe. "If we want to compete at the level of elite teams, we have to step up. But if I'm pushing you guys to do this, and it's my insanity, then we need to stop." Everyone agreed to keep working, then Gabe approached Amir afterward and said, "We need to do this."

On the bus to Long Beach, Amir also thought again about how his students were all rookies at a competition where teams had veteran drivers, veteran pit crews, and veteran scouts because they drew students from more than just the senior class. His team had built an incredible robot, had pushed the envelope of what was possible, but with all the randomness and luck involved in these competitions, that didn't mean they would win. He considered what defined success and how much his academy needed to do well in these competitions. He had long ago learned how challenging it was to attract attention for the academy. The football team might need to win a single regular-season game for everyone to go crazy for their sport. But with an educational activity like robotics, only extraordinary success would make people take notice.

The team had to make it to the Atlanta Championship, and for that they needed to win one of the two regional competitions they had entered that season—this Los Angeles Regional in Long Beach or the upcoming Sacramento Regional at UC Davis. Then Amir would be able to promote their success and, hopefully, raise the matching funds he needed for the grant before the deadline.

He was thinking about this and everything else when the bus rolled up to the Pacific Inn motel near Long Beach at 10:30 P.M., when his head hit the pillow in his room a half hour later, and when he awoke the next morning at the crack of dawn.

"Team 1717 is on the short list of teams capable of winning this event," Luke quoted from ChiefDelphi, which he checked several times a day. The rest of the team shoveled down their breakfast at the motel café. Amir downplayed the prediction. He had enough to worry about already, and the burden of his students expecting to win might put him over the edge. Soon enough, they were back on the bus, and on their way to Long Beach Arena, where they would have all of Thursday to work on their robots and participate in scrimmages with other teams. This was the practice day. Friday and Saturday were competition days.

Their driver, Roy, had brought an outdated map and soon found himself lost. They made a complete circle before ending up back at the motel. Amir wanted to be at the arena when the doors opened. They needed all the time they could get. He looked out the window, scanning for landmarks, while several students rummaged through their backpacks for their phones with GPS. A new route was mapped. They were back on track.

They arrived at Long Beach Arena a few minutes before officials allowed teams to enter. The structure's ten-story, circular exterior was painted to look like an ocean with gray whales frolicking in its waters, a strange juxtaposition to the event about to unfold within its walls. Students and mentors rushed inside.

Amir paused by the field where they would soon be competing. The 27-by-54-foot regolith field with its clear Lexan walls looked smaller than he expected. This was probably because of the breadth of the arena floor and the long stretches of stands that seated more than ten thousand people.

He then hurried past the massive black curtain that separated the field from the other side of the arena where "the pits" were located. There, large wooden crates papered with shipping notices and marked THIS SIDE UP were lined in perfect rows. Amir spotted the large black crate they had strapped their robot into almost four weeks before. He and his students fast-stepped it to their pit, a 10-by-15-foot area outlined in tape that would be their base of operations for the next three days. They removed the screws from the crate's side panel and pulled out their robot, nervous about the possibility that it had been damaged in transit. The PenguinBot was fine.

Aside from losing their way to the arena, Amir knew the first half of the day went far better than he could have expected. The team started straightaway on making some alterations to their robot. This involved fitting the wheel modules with larger pulleys to improve their gear ratios and swapping in refinements for their turreted shooter. They finished these changes quickly. At the end of their systems checks, Gabe said, "Every motor has been tested, and they all work!" They then passed inspections without a hitch, coming just under the weight limit.

In four years of *FIRST,* this was the first practice day Amir had found a spare moment to eat and take a bathroom break, let alone be finished with the PenguinBot in time to have his drivers participate in any of their scheduled practice matches. Stan, who had driven down on his own to Long Beach, was equally impressed with their early progress.

Chase was feeling none of their calm. Wearing jeans, a loose white T-shirt, and a knitted skullcap, he looked like he had just squared off against a ghost. He didn't seem to know what to do with his hands, first sticking them in his back pockets, then crossing his arms tight across his chest. "This is a big arena," he said, staring out at the field where he would soon be driving their robot. The matches that day were just scrimmages, and he was already a mess.

He circuited the pits to check out the competition and work off his nerves. The vast collection of machines made for the perfect diversion. One robot resembled a stepladder on wheels. Another had a circular wire basket on top that looked like it was built to collect radio signals. Some robots were made mostly of wood, some of aluminum frame and plastic netting, and some of Lexan and rolled metal. They were fitted with all kinds of mechanisms: conveyor belts, gyroscopes, rubber tubing, big expandable plastic bins, elevator shafts, and rollers and more rollers to pick up and shoot balls. There were three-, four-, and six-wheel drive trains, the majority tank.

Despite the diverse range of robots, they generally fell into three categories: shooters that were able to propel a single ball at a time; dumpers that stored a lot of balls in baskets that could be emptied all at once; and rectangular boxes that looked like they might, at best, be able to move across the field.

Seeing the competition made Chase feel slightly better. There was no doubt the D'Penguineers had one of the most complex robots with its omnidirectional drive system and vision-targeted turreted shooter. Amir and the other students were also feeling positive. The waves of mentors and students from other teams coming by the Team 1717 pit further encouraged them.

People were awed by their drive train, asking lots of questions about how it worked. Some pointed at the PenguinBot and just said, "Wow!" Many praised how incredible their robot looked with its anodized black body and gold hood. They took pictures of the PenguinBot with their cell phones like it was some kind of movie star.

Then came the practice matches. In their opening match, the PenguinBot didn't budge throughout the fifteen-second autonomous period, and when operated by Chase, it moved like a slug across the floor. In the second match, Chase had trouble collecting moon rocks, and Kevin kept missing clear shots. Once back in the pits, Amir said, "Everything needs to get way more efficient, or this is the way it's going to look for all our qualification matches tomorrow." After some adjustments to the drive train's traction control, they performed modestly better in the third practice match.

Afterward, Gabe approached Amir. He had an idea how to fix the automatic tracking software for their turreted shooter. "The camera's working fine in the pit, but it's not working in the field because of the lighting. It's—"

"Don't need an explanation," Amir said. "Just fix it. You're the man."

Gabe continued to detail what was wrong. Amir stopped him, held up his hand. He was confident in his student's abilities. "Nope, don't want to hear it. Heard enough. Moving on."

The game bewildered pilots Chase and Kevin. The field seemed like a confusing mess of robots. During matches, Amir stood behind them, urging them to collect balls when they couldn't even see their robot amid the mêlée. They mistook opponent trailers for those on their own alliance; they were running into jams of other robots; and they felt like the two-minute, fifteen-second matches were over before they had even started. Meanwhile, Turk and Bryan had switched off as shooters, neither proving terribly accurate in the scrimmages.

In their fourth practice match, their robot was dead on the field. It didn't move an inch the entire time. Outright panic gave way to relief when *FIRST* officials deemed the problem a "field fault"—their computers were merely having trouble connecting to the PenguinBot's controller. There was nothing mechanically wrong with their machine.

In autonomous mode in the fifth match, the robot shot across the field at maximum speed, crashed into the wall with a huge bang, went airborne, and then landed with a dull thud. Chase, Gabe, and Kevin looked at Amir like the world had ended.

"We might need to fix that," Amir said very calmly, given that the robot had almost broken the wall and killed someone.

Gabe worked on the problem while Amir loomed over his shoulder, asking how long until the autonomous routine was fixed. "You need to tell me, 'It's going to take a minute or thirty-nine seconds or—'"

"Thirty-nine seconds, Shaeer? Are you kidding?"

"You can do it."

Their frantic adjustments to the PenguinBot were reflected in the rest of the pits, which, by end of day, were a study in chaos, as sixty teams of high school students, all confined to narrow rows, all eager to use every power tool in their arsenal, hurried to ready their robots. Motors squealed, hammers banged, rivet guns popped, and voices strained in panic. And this was just the practice day.

At one point, another team called over the D'Penguineers to help fix their robot so that it could pass inspection. This was the kind of gracious professionalism that differentiated *FIRST* from other competitions. The pit crew also answered requests coming over the loudspeakers from teams desperately in need of spare parts or tools.

Gabe finished reprogramming the autonomous in time for their final practice match. The robot ran an effective routine, but the automatic tracking on the shooter was still not working properly.

Nonetheless, Amir was upbeat on the bus en route to dinner. "We're strategizing, we're focusing. Tomorrow, we're going to be in there with our flight suits on. I want us to be a team."

In a booth at California Pizza Kitchen, Gabe turned to Amir.

One could have packed for a weekend in the bags under Gabe's eyes. "If I don't say anything, I'm not going to sleep tonight."

"What's up?" Amir asked.

"That wasn't a field fault in the fourth match."

"They said our robot wasn't connecting."

"It was my fault," Gabe said. "When we were changing the code before the match, I took the code off but didn't put any back on. It just sat there because there was no code."

Amir smiled. "When did you realize it?"

"Soon as I looked."

"Just remember: Code is good."

The bus returned to the Pacific Inn after 10 P.M. In their room, Turk and Bryan joked about who would bring his best game as the team's shooter the next day. Amir had told them that they would be alternating matches, unless one was performing better than the other. Having practiced for the previous ten weeks, Turk didn't even want to think about not being the team's shooter.

Chase and Kevin, who were sharing a room, spoke again of the confusion on the field and wondered how they were going to figure out how to play the game in time. They shut off their lights early, needing rest for the big day ahead. Sleep came uneasily.

Amir crashed soon after returning to the motel. The practice matches had chipped away at his confidence in their robot, and the pressure he felt to win was intensifying every moment. No win, no publicity, no grant, no new building, no expanded academy. Slumber provided only temporary relief.

Gabe stayed awake until the early morning. There was nothing to do about programming now, but he needed to prepare for the team's presentation for the Chairman's Award. This was the most coveted prize in *FIRST*. It had nothing to do with winning matches; rather it was about how much a team embodied the standard of gracious professionalism and spread interest in science and technology in its community and beyond. Claiming the award, or its close sibling, the Engineering Inspiration Award, were the only two ways his team could earn a spot in Atlanta without winning a regional competition. Gabe was one of their lead presenters to the judges, and he had to get it right. It was their backup plan if their matches went south.

In their trademark black flight suits, Amir and his pilots, Chase and Kevin, huddled together outside the arena on Friday morning before the doors opened for the competition. They would not have another quiet moment before their qualification matches began.

"Slow down everything," Amir told his drivers. "You'll have tons of time out there, tons of balls and tons of opportunities. What will kill us is making bad decisions, then trying to make corrections. Our decisions need to be continuous and our actions need to be discrete. If we're going to go after a robot, we better well know that's the robot we want to go after. Things get so cluttered out there that I sometimes can't see what's going on. We need to be talking about where we want to go. It's all about visualizing what's happening—trying to see the future. It's about having a willingness to communicate.

"Last year, when our team was going for the ball, I told them to just *slow down.* Every time I forgot to say it, they screwed up. It wasn't their fault. It's high stress out there. You're trying to win, you're getting scored on. Slow and steady wins the race. Visualize yourself as like"—Amir switched to a singsong tone—"'I can go anywhere I want to go, and do everything with purpose and intent.' Other teams are going to be like, 'No, yes, go for that, no, do that, wait, no, yes.' We want to be on a way higher plane. And it's hard. There's only two minutes, you know?"

Chase and Kevin both nodded, so overwhelmed by the flurry

of words that they didn't recognize they had just been given a speech to remember for the rest of their lives.

When the doors opened, teams and their fans raced inside to get the best seats in the arena. With sixty teams, each with roughly sixteen to twenty students, plus mentors and parents and other spectators, there were more than two thousand people pouring inside. Amir hurried toward the pits, his drive team and pit crew unable to match his pace. He went straight to collect the match list to find out when they were competing, and who with and against.

FIRST competitions started with qualifying matches. At this regional, teams would have seven qualifiers. Each would be a three-on-three match (Red Alliance versus Blue Alliance). A computer selected the teams for each alliance by random. Luck dictated whether the D'Penguineers were paired with, or pitted against, strong or weak teams for each match. At the end of qualifications, the sixty teams were ranked by their number of wins.

The top eight teams would become alliance captains for the regional's elimination rounds. They would then create their own alliances by drafting two teams they wanted alongside them at the competition. The number-one-ranked team earned the right to make the first alliance-partner selection, the number-two-ranked team the second pick, and so forth.

A team could lose every match in the qualifying rounds and still be selected to be in the elimination tournament if it exhibited abilities that one of the eight alliance captains thought would prove useful. Still, a team's best shot at being crowned the regional champion was to finish the qualifying rounds among the top four teams so they could pick the best partners. Because there were only seven qualifying matches, a single loss likely meant that the D'Penguineers would not be among those teams. Every match counted.

Amir scanned the list for their team number. He found their first match and looked across to see who they were paired with: Teams 1644 and 691. Their opponents were Teams 702, 1382, and 2584. Right now these numbers were meaningless to Amir. He needed to know who these teams were, what their robots were capable of offensively and defensively, how good their human players were, how many balls they could score in a match, everything. This was why scouting teams throughout a competition was so important.

Andrew Hsu and Stuart Sherwin were in charge of scouting, and every D'Penguineer not working in the pits was responsible for watching matches and filling out data sheets on the fifty-nine other teams at the competition.

The pit crew rushed through one more systems check on the PenguinBot. Gabe manipulated the joysticks to test its motors. Everything was running fine. Amir and his two drivers hurried to meet with their alliance partners to talk strategy. Before they made much progress, an announcement came over the loudspeakers. "Let's get this started! The pits are now closed. Please take your seats for the opening ceremonies."

The D'Penguineers assembled in the stands overlooking the field. With their matching black flight suits, the students gave the appearance of a team for the first time. There were some serious speeches about the paramount importance of science and engineering. A taped video of Dean Kamen and Woodie Flowers played. Then the master of ceremonies vaulted onto the field. He wore a Krusty the Clown–style wig and a cape. Bunny ears flopped on his head, and his entire outfit was covered with hundreds of team pins. He revved up the stands with calls of "Are you ready?"

After an Air Force colonel sang "The Star-Spangled Banner," the qualifying rounds began. It wasn't long before Chase and

Kevin were wheeling their robot from the pit toward the playing field. Ahead of them, Amir and Turk cleared a path through
the scrum of robots, drivers, and their pit crews. Gabe remained
behind, confident that the day would go well. In his mind, they
had worked too hard on the robot for any other possibility.

In the narrow, crowded line to the side of the field where
teams waited on deck for their match, Amir gathered his alliance together. Because it was early in the competition, he didn't
have much scouting data on their opponents. His main worry
was Team 1382, from Brazil. Their robot was basically an elevator that rose and dumped balls, probably not a big scorer, but
their human player was a killer. "Stay away from that shooter,"
Amir warned.

The match before theirs ended, its winners raising their arms
overhead before they hurried onto the field to cart off their robots.
Kool & the Gang's "Celebration" played from the arena's speakers.
The song selections between matches ran a bizarre cross between
a wedding playlist and a running mix. The kids reveled in the
music and a massive projector screen above the playing field
showed them singing and swaying to the lyrics.

FIRST volunteers acted like traffic cops, waving teams on and
off the field. Chase and Turk placed their robot in position and attached its trailer. The drive teams on each alliance then moved
to their stations behind the clear walls on their respective sides
of the field. Turk was standing in a corner outside the field, a bin
of moon rocks by his side, ready to shoot over the wall into an
opposing trailer. The other five human players were similarly positioned around the field in front of opposing alliance trailers.

Turk had unzipped his flight suit, revealing his white 1717
basketball jersey with TURK emblazoned across the back. He rolled
his shoulders and stretched his neck from left to right like he was
about to go into the boxing ring.

The master of ceremonies jumped onto the field. He announced the competitors in the match, raising their respective flags overhead. "And from Goleta, California, Team 1717, the D'Penguineers!" he boomed, swinging their white-and-black flag.

Amir and his drive team performed their penguin dance, arms straight down by their sides, hands out perpendicular, hopping from left foot to right in a perfect waddle. Then they waved to their team in the stands. They would do this before every single match.

Settling back in line, Amir placed his arms around Chase and Kevin. "This is it. Focus."

"Blue team ready to go?" the emcee yelled to the opposing alliance. They nodded. Referees in black-and-white-striped jerseys moved into place on the sidelines.

With his narrow frame lost in his flight suit, Chase looked up into the rafters. He was still trying to get hold of his nerves. Their six-week build season, and all the practice and work afterward, would be defined by these seven two-minute-and-fifteen-second qualifying matches. Ten weeks against little more than a quarter hour.

"Red team is good to go?" the emcee said, turning to 1717 and their alliance partners. They nodded. Chase turned back to look at the field, his eyes wide as half-dollars.

"All right. Three, two, one, go!"

Los Angeles
Regional Competition
LONG BEACH ARENA, MARCH 12-14
PART II

Can there be any more drama?

—GABE RIVES-CORBETT

*T*he buzzer sounded, starting the first match of the season for the D'Penguineers. Chase and Kevin toed the white line separating them from their driver stations, preparing to leap for the controllers as soon as the autonomous period ended. In the far corner, the PenguinBot autonomously drove forward about 7 feet and then spun partly around. The prerecorded routine placed their robot between its hitched trailer and the nearest human player on the opposing alliance. By lofting several shots, their opponent still landed a couple of moon rocks into the PenguinBot trailer for four points. Turk more than made up for these by sinking four consecutive shots into an opposing trailer. He was off to a hot start. In the stands, his team cheered. "Turk! Turk! Turk!"

The six robots crowded the center of the field, their trailers jammed up against one another. Amir kept his hands on Chase and Kevin's shoulders to remind them not to step forward over the line until the fifteen-second autonomous period ended. But off to their left, an eager driver on their alliance did exactly that. Amir waved him away, but too late. A ten-point penalty.

Another buzzer signaled the end of autonomous. Chase and Kevin rushed forward and grabbed their joysticks to take control of their robot. The D'Penguineers alliance was down 12–8, even without the penalty. Two minutes remained in the match.

Chase shifted his joystick forward. The PenguinBot responded, barreling down the far side of the field. Kevin spun the intake roller to suck balls into the helix. They started the match with seven balls in their robot, but they were aiming to collect as big a load as possible before they made their first hit on the opposing alliance. In the corner, Chase struggled to maneuver the front of the robot onto a path of balls, even though they seemed to be everywhere on the field.

"It's already a more hectic match than any other we've seen," called the play-by-play announcer. Chase spun the PenguinBot about the field, trying to avoid clusters of other robots. The opposing alliance racked up several more points, mostly from the human player the scouts had told Amir to avoid.

The PenguinBot finally collected enough balls along the perimeter to go on the hunt. Chase pivoted the wheels 90 degrees, and their robot aimed toward the center of the field. The PenguinBot then turned to follow the elevator-like robot of Team 1382. Chase was lined up on their trailer, but Kevin could not get their shooter to lock onto the target.

"Fire! Fire!" Amir said, hovering behind his drivers.

Kevin switched to manual mode, but by then it was too late. Their opponent scooted away toward the side of the field. Chase followed the robot and pinned its trailer against the wall with the front of the PenguinBot. Kevin had a perfect line of sight for the shot. He rotated the turret left, then punched the fire button on his joystick. A torrent of balls unleashed from the hood, half filling the opposing trailer.

On the PA, the announcer said, "Team 1717 scores fast with their overhead turret system."

With one minute, twenty seconds left on the clock, the scoreboard put the D'Penguineers' alliance on top. Amir pointed several times toward the corner where there was a mass of balls to be collected. Off the PenguinBot went. Chase wasn't driving quickly, but the robot was going wherever they wanted. And none of the other robots on the field, whether on their alliance or their opposing one, had anywhere near their firepower. With thirty seconds left in the match, Chase tracked down another trailer and unloaded. All but a couple of shots missed.

The match was over. Win: 56–22.

Kevin and Chase clapped each other on the back, then sprang onto the field to remove their robot. Turk beamed. He had scored at least eight moon rocks for sixteen points, a big contribution.

After they returned the robot to the pits, Amir corralled his drive team and scouts. They broke down what they could have done better: Autonomous needed to spin their robot around farther to shield the trailer from the balls; Chase had to maneuver better while collecting balls; Kevin should be faster on the trigger; and their turret code required fixing. Foremost though, they recognized that the communication between Amir and his two drivers needed to be much more efficient. Often during the match Chase had no idea what Amir or Kevin wanted him to do, nor had they always been clear on where and why Chase was maneuvering the PenguinBot as he had.

Their next match was in less than half an hour. They were up against Beach City Robotics (Team 294), who Amir reminded his scouts were "notorious for owning us" in previous seasons. Stuart and Andrew predicted, "I think we're going to win it."

Gabe came running up to Amir. He and Yidi had just finished their Chairman's Award presentation. Gabe recounted his speech about how they were different from other teams: all seniors, all receiving class credit. He had admitted to the judges up front that they hadn't mentored any rookie *FIRST* teams or done much volunteer work, two key factors for the award, but then said the academy was doing so much more to spread the passion for science and engineering. Within Dos Pueblos High School, they had made robotics cool. The kids wore their flight suits before competitions just like football players did their jerseys. They were raising grants to expand the academy to reach more students, and one day their curriculum could be spread to schools across the country.

"They even let us go over the time limit," Gabe said. "We knocked it out of the park."

Amir hugged Gabe. They both exaggerated the embrace because it felt awkward, but it was heartfelt nonetheless.

Soon enough, Amir and the drive team were lifting their robot onto the field for their second match. Prior to the starting buzzer, one of their alliance's human players grabbed a moon rock to shoot. Ten-point penalty. Teams were still new to the competition, but this was absurd. Halfway through their autonomous routine, an opposing robot blocked the PenguinBot, leaving their trailer vulnerable again. When Chase took the drive controls, he was slow to collect balls where Amir directed him and ended up maneuvering into a traffic jam that persisted for over fifteen seconds.

Then a robot on their alliance stopped moving. It was dead on the field, an open target. Team 294's robot, Orange Force, moved in to take advantage. Amir saw this, but the PenguinBot was across the field, and they needed to focus on offense. He told their other alliance partner to block Orange Force before it could line up on the dead robot. Shell-shocked and obviously overwhelmed by the game, their coach just stared at Amir with what Amir would later call "crazy eyes."

"Can you get over there?" Amir pleaded.

The coach did not have his driver impede Orange Force, and by the match's midway point, the D'Penguineers' alliance was down by eighteen points.

Amir rocked back and forth, shifting his gaze between the scoreboard and the field. "We're losing. We're losing. You need to do whatever you can do," he urged his drivers. They managed to unload on one robot, but the opposing alliance, now ahead in the match, shut down the PenguinBot, pinning it between all three of their robots until time ran out.

Loss: 60–46.

"When a robot dies, it's over," Amir said afterward. But then he started to pick apart what he had done wrong. He had spent more time coaching his alliance partners than his own drivers, and Chase and Kevin were somewhat lost without him. He could have had Chase push their dead alliance robot to a corner, away from their opponents.

With the loss, their chance at a top ranking, no matter how well they did in the rest of qualifications, was slim.

Lunchtime offered Amir and his team more time to dwell on their defeat. While munching on Subway sandwiches by their bus, several students said they should have gone for a super cell. Their earlier strategizing had ignored the super cells, thinking that it took too much time to exchange empty cells for them, and the super cells could only be shot in the final twenty seconds of the match.

But the closeness of the scores they had seen so far at Long Beach was proving that the fifteen-point super cell could be a game changer.

"With one shot," Turk said, "we could've won that match."

"Yeah," Chase said, sitting cross-legged on the parking-lot pavement. "Woulda. Coulda. Shoulda."

Their lead scout, Andrew, said, "Other teams are going to study every second of that match and try to find a way to beat us."

Before their next qualifier, Gabe refined their autonomous routine so that the PenguinBot spun completely around to block incoming shots from the human player.

At the start of the third match, Turk rained balls down into a trailer that stopped in front of his post during autonomous. He was at the center-line position, where the human player had to remain belted down to a seat. It was an awkward position to throw balls from, but he was still on target. Their alliance human players were also shooting well from their standing positions in the corners. When the driver-operated period began, their alliance already had fifty points on the scoreboard.

One major score by the PenguinBot ran up the advantage even higher. Two robots pinned the PenguinBot against the wall, but Chase managed to free it by taking advantage of a slender gap between the two robots. Their swerve drive made that possible.

Win: 88–6.

"Fabulous job!" Stan cheered from the sidelines, unusually emotional.

Amir went in for a hug, but Stan put up his hand. "No time."

Jamil Abo-Shaeer, who had flown in from the East Coast to support his brother, said with a dry smile, "There's a real lack of consistency here. Ten minutes ago, it was like it was all over. Now, wow. The highs are high and the lows low."

"I don't think it matters about being number one now," Stuart said, pushing back his long hair from his face. "Now it's showing whoever's in that spot that we're the one to pick."

Chase was silent, thinking of the match ahead. He could

scarcely remember what had just happened. "My mind is two minutes and fifteen seconds long right now."

"You're like the guy in *Memento*," a teammate said.

"I need to get tattoos on my body that say, DRIVE TRAIN, COLLECT BALLS, SHOOT BALLS, and LISTEN TO SHAEER."

The D'Penguineers also dominated the fourth match. Their opponents pinned them down again, but this time they already had a full helix of balls. Thanks to the PenguinBot's rotating turret, their opponents paid the price of playing defense against them. Watching the shooter launch, Gabe said, "It's not even funny how much I love that shooter."

Win: 74–36.

Gabe told Chase after the match, "I heard another team say that they were screwed having to go up against us."

Their fifth and final match of the day proved ugly. There was a problem with the shooter continuing to propel balls after Kevin shut down the flywheel, wasting shots. Chase was also pinned for a quarter of the match. Still, they won 80–35 and finished the first day of qualifications ranked fifth.

Before the pits closed at 6 P.M., Gabe ran through a systems check, identifying the problem with the shooter: The kicker wheel kept running after the flywheel stopped. It was a simple code adjustment.

Amir was off speaking to the mentors of the top-ranked teams, seeing who they were thinking of picking during alliance selections. They told him that 1717 was high on their lists but made no guarantees. This kind of politicking was common practice in *FIRST*, and it became even more heated right before selections.

Once everyone was on the bus, Amir took the microphone. His students were rowdy and still charged by their three consecutive victories. He felt they needed to understand the challenges they would face the next day.

"Listen up," Amir said, his voice weak. "You need to know how consequential our ranking is. For us, our fate's already written."

"Awwww," the students said, deflated.

As the bus rolled away from the arena, he explained that two of the top three ranked teams, 1388 and 399, would likely pick each other, creating a very tough alliance to beat. Even if the D'Penguineers won their two remaining qualification matches on Saturday, they would probably not be ranked any higher than fifth. That left them at a disadvantage during alliance selections, and it was possible they could be paired with less-than-ideal partners for the tournament.

The more Amir spoke, the more dispirited his team became. He was tired and overwhelmed. The day had been a feverish rush: strategizing with alliance partners, coaching his rookie drivers in a game that felt to him like navigating in a blender, making sure the robot was in good shape, and keeping the scouts focused. All the while he had felt, and continued to feel, intense pressure to see their team win, particularly for his students who had worked so hard and, knowing they only had this one season, wanted so much to win.

Amir hesitated, sensing the effect his words were having. "When I say our fate's sealed, that's just about who's going to pick us, not whether we're able to win or not win. I don't want you to think that. But you have to realize that the success is in what we've created, the fun we've had, the bonds that we've made between each other—all that stuff. I heard this quote. It was 'Success is for the moment, and only that moment.' The point being that you're going to succeed in that instant, and you're going to have that feeling of success, but what are you going to have after that? After that, it's the experience of the whole entire package, and that's what you're going to look back at. Tomorrow, we're going to do the best we can, and whatever happens after that happens after that."

As he sat back down and looked out the window into the dark, he wondered if he'd been too blunt. He knew the team needed to believe they could win, but he also thought they should know all the facts, good and bad, about their situation. He didn't know how to balance the two.

On Saturday morning, when Amir opened his motel-room door, he found an empty milk gallon at his feet. The attached note read: "I did it—51 minutes!"

He laughed, his spirits lightened after a long night of worry. As soon as he saw his drivers downstairs, he knew he'd found the culprits. The previous year, against Amir's explicit request that they not do so, some students had tried to drink a gallon of milk in an hour, a feat that was impossible and induced a lot of vomiting. Amir had made it clear this year that he wanted no repeat attempts, so, of course, they had made a joke of it by emptying the milk down a drain and leaving the gallon jug at his door.

On the bus, Gabe sat down next to Amir. After the sobering talk by Amir the night before, the programmer had been thinking of any possible changes he could make to improve their chances. He told Amir he had been mulling over code for the shooter that would provide a mode on Kevin's joystick that automatically took the highest-arc shots. Since their shooter's tracking system was not working well, this mode provided an override. Amir agreed that they should implement the code that morning.

Gabe also reminded Amir of how several times Chase had been pinned against walls by robots that came at them from an angle. Gabe offered a solution: force vectors.

If a robot was pinning the PenguinBot at an angle, Chase should not try to break free by pushing their robot straight back against their opponent or trying to run along the wall. Instead he needed to take advantage of the sideways force exerted on the

PenguinBot by the pinning robot by escaping diagonally from the wall. It was just like a basketball thrown against a wall. It doesn't bounce straight back but rather moves away at the reflected angle from which it was thrown.

Amir smiled. He had been thinking the same exact thing, and it was doubly nice that the physics he had taught his students had stuck.

As soon as they arrived at the arena, the team brought the PenguinBot to the small practice field in the pits. Gabe and Amir showed Chase what they wanted to do about the pins.

Then they also worked on an idea Amir had. Often when pinned against the walls, one side of the PenguinBot was on the carpet, which had more traction than the regolith. The carpet allowed them to apply more torque to these wheels without them slipping. Amir wanted a traction-control mode that would provide more power to the wheels on the carpet than those on the regolith, giving them the ability to push out of pins more easily. Depending on which side of the robot was on the carpet, Chase would hit one of two buttons on his joystick to switch into this traction-control mode.

With Amir pushing against the frame of the robot, they tested increasing the power they gave the motors until the wheels began to slip. Twenty minutes later, they had a mode that provided 25 percent more pushing power.

Then Amir hurried through the pits to find his scouts and his two shooters, Bryan and Turk.

"Guys, we need one human player. A mentor came up to me worried that we have two shooters. It doesn't show consistency." He turned to his lead scouts, Andrew and Stuart. "Who's making more shots?"

It was even, they said, though Turk was better at the seated position at the center line.

"I think you guys need to decide," Amir said.

Turk stared at the floor. He wanted to be the team's shooter. He had practiced for almost two months. Finally he looked up at Bryan, *please* written all over his face.

Bryan glanced away. He knew Turk had never had the chance to realize his dream of playing high school basketball. Finally, Bryan said, "Turk's better at shooting seated. I think he should be the one."

It took every ounce of Turk's willpower not to raise his fist in the air and jump up and down.

"I'm going with you then, Turk," Amir said.

Turk smiled, then thanked Bryan. Over the loudspeakers came the announcement that the pits were closing for opening ceremonies. Amir retreated to a picnic table outside the arena to talk with Stan. While Amir was more focused on overall coaching, his lead mentor was in charge of the robot and pit crew. They ran through the changes they had made on the PenguinBot, agreeing they would help, though Stan was worried about last-minute code changes without testing.

"Are you having a good time, Stan?" Amir asked.

"I'm having an acceptable time." He smiled thinly.

"My wife told me you said you were having a good time, but you won't admit it to me."

"No, your wife is all about the feelings, and so are you, so I just try to tell her what she wants to hear. Same as you." Then, almost in the same breath, Stan said, "We're missing collecting a lot of balls, and the way Chase is driving, slow mostly, doesn't say 'I'm in charge.'"

Amir nodded. "At times our robot looks like an underwater pool sweeper that glides and looks cool but doesn't freakin' fly."

Back in the arena, Amir told Chase he needed to use less finesse and more speed, ramming into robots if need be. They needed to look good on the field in the last two matches so teams would want them on their alliance.

When Chase and Turk lifted their robot onto the field for their sixth qualifying match, the team chanted from the stands. "Turk-a-licious! Turk-a-licious!" During the match, Chase proved much more aggressive, showing for the first time how fast the PenguinBot could be. Their alliance was leading with thirty seconds left when a robot collided into the PenguinBot and it died.

Chase and Kevin shifted their joysticks left and right. Nothing.

They lost 64–48, ensuring that they would not finish among the top eight teams.

When they went to pull the PenguinBot off the field, they saw that the battery cable had come unplugged. Amir had told his mentors and students that the cables needed to be zip-tied before each match, but they hadn't done it. He made it very clear now that he didn't want this kind of stupid mistake again.

Amir walked off, worried that some of the top teams might think their robot was an unreliable pick for the tournament. But he was still confident his team would be picked in the alliance selections. The question now was whether or not it would be by one of the first two or three teams.

Winning their final match 76–34 calmed Amir slightly. They finished twelfth out of sixty teams. Respectable, but not what anyone on the team had expected. Now everything was about alliance selections.

"This is it," Amir said as representatives from the eight alliance captains filed onto the field to make their picks.

Team 1388 was ranked first. Team 597 second. Team 399 third. Amir met with Stuart, who would go onto the field if and when someone selected the D'Penguineers. Stuart had a bunch of folders with scouting data on every team at the competition, including his own, tucked underneath his arm. He would be responsible for helping choose the third team for their alliance.

"We need a team who can pin but also score a little," Amir said.

The blue-haired emcee strode onto the field, made a few comments about the process, and then handed the microphone to the captain of the top-ranked team. In a squeaky voice, the student said, "Team 1388 would like to extend an alliance to 399." That team's captain walked over and said, "Team 399 accepts." Hands were shook, and cheers rose from the respective teams. No matter their third partner, that would be a tough alliance to beat.

Then it was the number-two-ranked team's turn. Many thought the Wolverines (Team 597) had gone undefeated because they had easy qualification matches. Their robot could store thirteen balls and was a mix between a shooter and dumper. The students were from an inner-city high school in Los Angeles.

Without hesitation their captain said, "Team 597 would like to extend an alliance to Team 1717."

In the stands, the D'Penguineers hollered and whooped. Standing up in the bleachers behind them, four bare-chested students hoping their team would be the next one selected chanted their team number. On their pale white skin in blue paint was written "5-8-9!"

Stuart walked over to the emcee with a slow, loping stride. "Team 1717 accepts." It was not ideal, but with 1388 and 399 together, there were not a lot of better alternatives. Being picked in the number-two alliance also had its advantages. Given how the tournament brackets were organized, their alliance wouldn't have to face the number-one alliance until the regional finals—if either of them advanced that far.

In the draft for their third alliance partner, Stuart urged their alliance captain to pick Beach Cities Robotics (Team 294) because of their defensive abilities and record of past success. Done.

The alliance convened for a brief strategy session, then everybody broke for lunch. Before the start of the quarterfinals, Amir and his drive team spent most of their time at the pit practice field running drills to collect and shoot balls.

An announcer called 1717 for its first match. Amir and his drivers continued their "drill and kill" session for a couple of minutes, then rushed the robot toward the queue for the tournament where Turk was waiting for them.

Before their first quarterfinal match, Amir strategized again with their alliance partners. "Collect and score as much as possible. Go for the best robot, keep them locked down. We'll trawl and hunt."

Chase stared straight forward. Stuart and Andrew, scouting clipboards in hand, tried to relay some information to him about their opponents, but he waved them away. "I can't hear this right now." He was trying to force everything out of his mind except what he had to do to drive the robot.

Amir paced back and forth, hands on his hips, fists balled at his sides. Kevin tapped his hands on his legs, muttering again and again, "Now it really matters." Turk bounced up and down, throwing his arms across his chest and behind his back, stretching himself out.

The drivers and coaches from the twenty-four teams (eight alliances of three teams) in the tournament crowded the lanes around the field, moving their robots, running through systems checks, and huddling to talk. Like an announcer at a prize fight, the emcee introduced each team in the quarterfinals. On mention of their teams, the drivers and coaches pumped their fists and clapped. In the stands, costumed mascots led their teams in singing fight songs or chanting their team numbers. Between matches, music roared from the loudspeakers and students danced in the aisles and in front of the field. It was getting wild.

Seconds before the match, Amir nudged Chase and Kevin. "Are you okay?" They both shook their heads, both looking very much *not* okay.

Each round of a *FIRST* tournament was set up in a best-of-three match format. The first alliance to win two matches advanced to the next round. The other alliance was eliminated. There was no more room for errors.

In the quarterfinals, Chase drove the robot as well as he ever had, making their alliance all but unstoppable. He collected balls efficiently, maneuvered swiftly, and avoided most pins. The one time an opponent locked the PenguinBot down against the wall, Chase used Gabe's vector move and freed it. Kevin was fast and accurate with the turreted shooter, even though their automatic tracking system was still not working. Turk showed that he was as important an offensive weapon as any of their alliance's robots. Of the thirteen balls in his bin, he averaged at least seven or eight shots. With strong defense from their alliance partners, they sailed through the quarterfinals, winning the first match 86–30, the second 102–34.

On to the semifinals.

They were up against the NERDS (the Nifty Engineering Robotics Design Squad, Team 1726), who had gone 6–1 in the qualifications. They had a helix/turreted shooter similar to the PenguinBot's. The NERDS were allied with the Beach 'Bots (Team 330), former champions and consistently one of the best teams on the West Coast. The D'Penguineers had beaten them in the finals of the 2008 San Diego Regional, but in the year prior, the Beach 'Bots had twice eliminated them. Their third partner, the ThunderBots (Team 980), were no pushovers either.

"We can do this," Amir said, all three drive teams circled around him. The other two coaches stood in the background. "The NERDS are our main target. Also pin the Beach 'Bots if you can. We need to hit their robots."

In the stands, the D'Penguineers were on their feet, screaming "Red Alliance!" as the buzzer sounded. The autonomous period was a draw. As soon as the drivers took control, the BeachBot and another opposing robot zeroed in on the PenguinBot, getting in its way and trying to position for a pin. Kevin scored the seven balls preloaded into the helix, but Chase couldn't do much collecting, as their robot was pestered everywhere it went.

"Don't lose your patience!" Gabe shouted from the stands. As the seconds ticked down, the score remained tight.

Chase managed to break free and home in on an opposing trailer for a few more points. Something was off on the shooter. The adjustable hood was angled too steeply, and balls were bouncing off the top of their own robot. It was a low-scoring and close match.

At the buzzer, the scoreboard read 50–48, but the score on the projection screen was an unofficial count based on the number of balls that *FIRST* officials had counted in real time with handheld electronic clickers. Often this was imperfect because robots could shoot faster than the officials could click—or they just missed shots altogether. After each match, officials swarmed the field, pulling balls out of trailers and placing them into bins for a final count. A minute passed. Techno music played in the background, raising the level of tension.

Gabe hurried down to the field and high-fived Chase and Kevin. "Come on. Let's see it," he said, unable to stand still.

"Here it comes! What a squeaker!" the emcee announced. "Fifty-eight to fifty-four. Red Alliance takes semi number one. Barely."

Amir finally exhaled and met with his alliance before the next match. The coach of the Wolverines told his drivers that they needed to help out 1717 because everyone was chasing them. Chase stood close to Amir throughout the rushed conversation, as if his teacher's presence instilled him with confidence.

In the second semifinal, after a low-scoring autonomous for both teams, the PenguinBot was free early on to make a clear shot on the BeachBot's trailer, but most of the balls missed their target as the opposing robot spun away. Almost immediately they lined up on the PenguinBot's trailer.

"Don't let them dump on you!" Gabe warned from the stands. Just as the BeachBot released its balls, Chase sped away, with no more than a couple of balls dropping into their trailer. He swerved around the field, collecting balls, while the Wolverines had trouble pinning down any of their opponents. Only their partner robot Orange Force was scoring a fair number of shots. At the midway point in the match, the real-time score was tied.

Chase managed a couple more small scores but looped around the field somewhat aimlessly. Super cells were in play for one of the first times in their matches. Both alliances had exchanged an empty cell for a fifteen-point super cell, but both missed scoring them in the final seconds.

When the match ended, the scoreboard had the alliances eight points apart.

"That was the craziest match," Amir said. "You think we're in?"

"Ugly," Chase said. "That was ugly."

"I got trigger-happy," Kevin said.

The announcement: "Sixty-four to fifty-four. Red Alliance."

It wasn't a pretty win, but they had made it to the regional finals.

Amir headed over to the opposing alliance to shake their hands and congratulate them on their competition. A NERDS mentor told him, "Your robot shooter is *so* good."

Chase and Amir stood by the side of the field together, overwhelmed by the moment, not speaking. They watched the third match of the other semifinal, expecting the top-seeded alliance that Amir feared so much to win. They lost.

The D'Penguineers would now face the RoboWarriors (Team 2659), the Metalcrafters (Team 207), and the Greybots (Team 973), whose bright orange robot, the Raptor, maneuvered the field with lightning speed and had a very quick dumper. This alliance had worked together effectively in the earlier rounds, and they had few weaknesses to exploit.

FIRST officials prepared the field while the Village People's "YMCA" blared from the loudspeakers. The arena stands erupted with dancing high school students. Amir huddled close with the drive teams and his scouts so that they could hear one another. Two more wins and they would be the regional champions.

"What do I need to know?" Amir asked.

Andrew and Stuart started speaking at the same time, not making much sense. Stuart was particularly nervous, his safety glasses so fogged up with sweat that he couldn't see. Amir stopped, took a deep breath, closed his eyes. "Guys, we don't have much time. We need to be cool and focus."

Stuart cleared his glasses, forced himself to speak slowly. "The RoboWarriors have an amazing pinning robot. We want to send Orange Force against them and send Wolvie after the Raptor. That puts us one-on-one with the Metalcrafters, who have a big dumper. Also, this alliance has a curly-haired freshman who's a great human player. Stay away from him."

Meanwhile, Gabe circled around the robot with a flashlight, looking at every single connection. "Everything looks good," he said.

A delay on the field allowed Amir to bring Turk, Chase, and Kevin into a hallway leading into the arena to escape the commotion. "The Electric Slide" had followed "YMCA," and the place was going crazy. Trying to catch his breath, Chase bent over and placed his hands on his knees.

"It's man to man now," Amir said. "We have to deal with the Metalcrafters. Don't follow around other robots. Avoid the Greybot like the plague. Stay in the middle of the field, away from the corners, where we can get pinned. And be calm. We get balls. We shoot. Everything's good."

They reentered the arena to Queen's "We Will Rock You." The stands were alive with students stomping and singing.

They positioned their robot on the field. Kevin wrapped his arms around the shoulders of Chase and Turk. "We can do this."

The emcee jumped onto the field. "Are you guys ready for the finals? Is it going to be Red or Blue?" He pointed left. "Red?" Then right. "Blue?" Then quicker: "Red? Blue? Red? Blue?" He threw his hands high above his head, then hustled off the field for the match to begin.

After the bell rang for autonomous, Turk sank six balls into an opposing trailer. Their alliance was up 22–6 by the time Chase and the other drivers stepped forward. The Raptor made a huge score right after the bell because it was already lined up on a trailer. The score was almost tied.

From the stands, Gabe watched the match with the rest of the team and a bunch of academy underclassmen who had come down by bus that morning. Together, they banged their black 1717 Thunderstix (inflatable plastic noisemakers). One student chewed on her black-and-white pom-poms as the score remained almost tied with sixty seconds left.

It was obvious how much more competitive the matches in the tournament were than in qualifications. Everything moved much faster. There were fewer clusters, more pins, and far more targeted scores.

With thirty seconds left, the PenguinBot hit an opposing trailer with a devastating number of balls, its second good unload. The scoreboard ticked up in their favor as the final seconds ran down. At the buzzer, Andrew flexed his arms over his head and released a guttural battle cry. The veins in his neck popped. Everyone else jumped up and down. The real-time scoreboard had their alliance winning 64–54. The D'Penguineers were now a single match away from victory.

But then there was a delay announcing the final score while the officials did a hand check.

Gabe made his way down to the field to do a systems check on the robot. Over the roar of AC/DC's "Thunderstruck," Amir said the delay might be a penalty instead of the count.

"For what? On who?" Gabe asked.

"I think one of our partners made a foot fault during autonomous," Amir said.

The match was close enough that any penalty might reverse the outcome. He and Gabe both looked toward the officials and asked, "What's the score?"

Finally the announcer indicated that there was a deciding penalty, then the scoreboard flashed the final score: 58–56. The D'Penguineers had lost because their alliance partner stepped over the line toward his driver station before autonomous ended.

"A very tough penalty," the announcer continued. "That changed the outcome."

Without the ten-point deduction, the D'Penguineers would have won 66–58. Instead of being one win away from being regional champions and advancing to the *FIRST* Championship in Atlanta, they now needed to win two in a row because of a foot fault.

There was no time to wallow. The pit crew swapped in a new battery on the PenguinBot, and Gabe ran a quick systems check.

The drive teams from the two alliances circled the field to shake hands as was *FIRST* custom during the final.

Amir returned to their alliance position, his back straight as a rod, his lips pursed. He wrapped his arms around his drivers and told them a stupid mistake had cost them the last match, but they needed to forget it. They could still win.

The buzzer sounded to start the second match. Once again, Turk hit his first eight shots into a trailer, giving their alliance an early lead.

Halfway through the match, the score was tied again and the PenguinBot found itself pinned against the wall by the RoboWarriors.

"Get out of there!" their teammates yelled from the stands before starting a chant "Seventeen-seventeen! Seventeen-seventeen!"

"Chase, throw it in carpet mode and start testing angles!" Amir shouted. One side of the robot's wheels was on the perimeter of carpet. Twenty seconds had passed. "We know there's an angle to get out."

Chase hit the button that increased the amount of power to the motors, the routine Gabe had programmed that morning. Then he shifted his joystick in several directions, looking for a way out.

Thirty seconds passed: still pinned. Forty seconds.

Then he found his angle. The robot pushed its way free, then spun around.

"Come on, Chase!" Amir shouted. "Find a trailer!" He and Kevin spotted a target in the center of the field and pointed toward it.

Chase looked at them. "I got it. I got it," he said.

He drove to the trailer but then kept moving right past it.

"Where are you going?" they yelled, not understanding why Chase had continued past.

They had six balls in their helix. They needed to score. With four seconds left, Chase cornered an opposing robot.

"Kevin," Amir said, "you have to take the shot. Go! Go! Go!"

Moon rocks streamed into the trailer just before the final horn. The stands erupted.

Survivor's "Eye of the Tiger" boomed from the loudspeakers as the *FIRST* officials tallied the final score.

"Can there be any more drama?" Gabe asked, rubbing the back of his neck, then his eyes.

The emcee declared, "Seventy to fifty-eight. We play a tie-breaker."

In the stands, the D'Penguineers screamed and hurled themselves into embraces. Even their bus driver, Roy, a former cop who had first met the team on Wednesday, cheered with abandon.

On the field, Kevin and Chase fist-bumped. Then Chase hugged Turk and anybody on the pit crew he could find. Amir hopped and skipped on the sidelines, reveling in escaping that pin. He was sure no other robot on the field would have been able to do it. He was smiling so broadly that Emily would later say, "I think that's the happiest I've ever seen him."

For a moment the arena quieted, then the theme from *Rocky* played on the speakers.

"Welcome to the main event," the emcee said. "Winner of this match wins it all."

When he introduced 1717 among the final teams, Amir and his drivers did their penguin dance one last time and waved into the stands. There, all of the D'Penguineers were clapping and chanting for their alliance, but their faces were straight and serious.

The match began with another resounding lead for the D'Penguineers. Turk was in the zone, and they led 28–0 on the scoreboard when autonomous ended. A quick dump by the PenguinBot stretched this lead even further.

Turk, who was out of balls to shoot, looked at the scoreboard and thought, *We might just pull this off after all.*

Then the opposing alliance roared back to life. The Raptor made two big dumps, and the score was almost tied with thirty seconds left. In the stands, Gabe and the others were silent. Nobody was clapping.

In the corner in front of Chase, an opposing robot crashed hard into the side of the PenguinBot where their electrical board was located. The smash was almost deafening.

"Look at the turret!" Kevin yelled.

The turret was randomly swinging back and forth.

"Whoa, dude," Amir said, keeping Kevin from hitting the "kill" switch, which would have disabled the robot for the rest of the match. The wheels started to rotate wildly. The helix was spinning. All without a touch of their joysticks. Every mechanism looked like it had lost its mind, and the whole robot was beginning to rattle and shake like it was about to break apart from the inside out.

"Oh my God!" Amir gasped. "Kill it. We're done."

Kevin hit the switch. The robot was dead on the field.

Their opponents took advantage. A blur of orange, the Raptor swooped down on the PenguinBot in the corner. Chase looked on as two balls dropped into their trailer, the shots that determined their fate.

The scoreboard flashed: 68–64. The D'Penguineers had lost.

In the stands, everyone was stunned. The other alliance celebrated on the field, hugging and raising their fists into the air, as Queen's "We Are the Champions" blared from the loudspeakers.

Then Gabe started clapping, wanting to be a good sport. Everyone joined him. "There's still Chairman's and the Sacramento Regional. We can get to Atlanta. It's not over."

Gabe hurried down to the field and followed Amir and the

drive team to the pits. They had already identified the problem. When they had been hit, one of the power connectors had popped out of the electrical board. This rendered many of their sensors dead.

"Removing that wire is like cutting someone's brains out and stomping them on the ground," Amir said. "We need to do a full-systems check now so we can assess the robot before we put it back in the crate."

They couldn't arrive at their next regional not knowing if some mechanism on their robot had been broken while the PenguinBot went crazy on the field. Gabe ran through the systems check, and everything looked okay.

They returned to the stands before the final ceremonies. Amir leaned in toward Emily, his whole body listless.

"That was awesome," Emily said encouragingly.

"The Greybots were scrappy like we were last year," Amir said, taking off his glasses and rubbing the back of his head. "We couldn't do anything else."

As the ceremony began, Amir still had faith that they might win the Chairman's Award or its sibling, the Engineering Inspiration Award. Not only would these awards affirm what the academy was doing, but they would also earn them a spot at the *FIRST* Championship.

They did not win either award.

Afterward, they disassembled their pit, packed up their bins and toolboxes, and finally lifted the robot back into its crate. The next time they would see the PenguinBot was at the Sacramento Regional in two weeks, their last chance to make it to Atlanta.

The Competition

*T*hat same weekend, halfway across the country, the **Heroes of** Tomorrow (HOT) found the success that had eluded the D'Penguineers. It was a surprising start for the General Motors–sponsored team from Detroit, given that only days before they had been forced to fundamentally redesign their machine.

The build season had started so well for high school senior Nick Orlando and his HOT team. On the eve of kickoff, he was driving home overnight in his family's Chevy van from their annual Florida vacation. He couldn't wait to find out about this year's game. They arrived in Milford, Michigan, late in the morning, a couple of hours past the kickoff, but Nick refused to call anybody to get the scoop. He didn't let his twin brother, Matt, or younger brother, Alex, both also on the robotics team, call either. Nick wanted to see the challenge for himself.

Once through the door of his house, Nick ran upstairs, his brothers in tow. His bedroom was a mess: clothes everywhere but in his closet; bed unmade; backpack spilling books out onto the floor; dresser top piled high with movie tickets, receipts, loose change, and other detritus from his front pocket; and all manner of electronics, the key pieces being a PlayStation 3, a 21-inch LCD, and his laptop. Sticking out of the closet was a box overflowing with speaker wires, power cords, AC-to-DC converters, and cables, including a big twisted mess of FireWires, USBs, and the like.

Nick fired up his laptop, the anticipation almost too much to bear. On the screen, the NASA simulcast began to play. Over the next forty-eight hours, fueled by stacks of Mountain Dew, Nick designed a three-dimensional, computer-animated version of the robot he thought would best meet the challenges of Lunacy. Balls would come through a narrow opening into a low basket. They would feed up a column, propelled by spinning loops of thin surgical tubing, into the shooter. There, a baseball-pitching-machine-like mechanism would force out the balls.

Nick presented the design to his team at their headquarters in Building 40 at the General Motors Proving Grounds, the site where the automaker test-drove its cars. The team liked the concept. Nick was pleased. The son of an airplane mechanic, he had always loved to tinker and build things. He had joined HOT as a freshman, so eager that he had proposed building a hovercraft. Now a fourth-year member of the team and its lead driver, he had not mellowed much. He considered robotics equal to any varsity sport at his high school, even if his classmates didn't, and he pursued it with equal zeal.

With more than forty students, a dozen mentors, and thirteen years of *FIRST* experience, HOT was a very strong team. They were former world champions and would have been so again the previous year if not for their defeat by the Ford-sponsored Thunder-Chickens in the Atlanta finals. HOT expected to be among the top teams every year, and this season was no different.

Working off Nick's concept and integrating other ideas, they set about building their robot. They stuck to their tried-and-true six-wheel tank drive train. They designed a single-ball-wide tower fitted with rotating bands of surgical tubing that delivered balls up to a turreted shooter. Balls were stored within the Lexan outer walls of the robot until they were fed up to the tower.

They fabricated their parts in an industrial-sized machine shop used by GM at the Proving Grounds. They even had the use of a computer-controlled water jet that cut sheet metal with bursts of water mixed with sand. All but one of their mentors were GM engineers. As the ship date approached, they had completed two almost identical robots, one for practice and the other for competition. They didn't have to rush or work long hours to finish either.

But there was a problem. In their early strategy meetings, they had figured that to be effective in the game their robot would need to be able to hold fifteen balls and fire them at a rate of two balls per second. The robot they had built was never able to hold more than eight balls or sustain that firing rate because balls kept jamming as they made their way into the bottom of the narrow tower. No matter what they tried to improve the funneling of balls into this tower, it failed. Finally, on February 17, they had to call it quits and seal up their robot.

Three days after the end of build season, mentor Adam Freeman, the team's new drive coach and one of their lead designers, sent everyone a note. "I feel we're at somewhat of a crossroads with our robot design. Many of the elite teams in *FIRST* have released pictures and videos of their machines doing amazing feats like shooting or dumping huge amounts of balls in a little amount of time. These teams are just as good at scouting, programming, and strategy as we are . . . and at this point their machines have more potential."

He then proposed that they come up with a new design to fix the bottleneck in the tower. According to the rules, teams were allowed to replace up to 40 pounds of parts on their machines during the eight-hour window they had to work on their robots before competition. Adam suggested that they use that window to overhaul their robot completely.

Nick jumped at the idea. He spent the next two weeks running through more ideas. His best one involved a mechanism for the hopper that worked like an auger, agitating the balls until they fed into the tower. He roughed out the mechanism on the practice robot, using pieces of Lexan to push the balls around. He was sure that they should run with it.

Then on Friday, March 5, a week before their first competition, Jim Meyer, a GM engineer and the team's other chief designer, had a different idea. Jim was raised on a small farm in Nebraska, and as one of his fellow mentors said, he was so good at his job that he could bang a piece of mangled metal with a hammer a few times and create a perfectly finished part. His idea was to double the width of their tower and shooter, creating a power dumper. They would have to eliminate their turret, but this design would solve the bottleneck problem and realize their ideal ball capacity and firing rate. Nick reluctantly relented on his own idea.

That Sunday in a frantic build session at the Proving Grounds, the team stripped their practice robot of its tower and shooter and machined parts to expand them for a two-ball width. The new shooter and tower was under the 40-pound limit.

Teams in Michigan didn't have to ship their robots to their competitions like the D'Penguineers had. Robots could be sealed in a bag at their build site, which HOT had done. Their eight-hour window to work on their robot could be used at their build site before they came to competition. On Monday, a designated official unsealed the bag. With the clock ticking, HOT disassembled everything on their competition robot but the chassis and drive train. They fitted in the new parts and rewired everything, and on the first test run, the robot operated with no jams, a fifteen-ball capacity, and a devastating fire rate. It was transformed.

On Thursday, March 12, sitting shotgun in a Cadillac STS heading into downtown Detroit, Nick had no idea how this first competition would go. Everything about it seemed last-minute, even the drive team's hastily arranged ride to their practice matches from a mentor who had the Cadillac out on a loaner.

Nick had driven their new robot only a couple of times, and he didn't know what to think of the final design. He was sure a turret was key to this game, but he was surer still that those eight hours transforming the robot would define his team more than the six-week build season itself.

The practice matches went better than Nick had hoped. People came up to him in the pits that night, commenting on what an incredible robot HOT had built yet again. If they only knew the whole story, Nick said to himself.

The true test of their new design came the next day in their first qualification match against their big rival, the Thunder-Chickens. "Let's do the best we can," Adam said to Nick before the buzzer sounded to start the match. "The machine's good. Let's worry about our strategy."

HOT beat the ThunderChickens by an overwhelming margin, and Nick never worried about their robot again. HOT went undefeated during the next two days and won the competition (allied in the elimination tournament with none other than the ThunderChickens). The following weekend, they went 18–0 for their second Michigan title, earning their robot a reputation as one of the most dominant offensive weapons in *FIRST*.

The perseverance to continue to improve their robot was a quality that Amir knew his own team needed to maintain if they hoped to go up against the likes of HOT in Atlanta.

Sacramento
Regional Competition

THE PAVILION, UNIVERSITY OF CALIFORNIA, DAVIS, MARCH 26-28, PART I

Don't worry, Shaeer. When we're out there, I'm scared out of my mind.

—CHASE BUCHANAN

On March 20, when everyone on the D'Penguineers checked their individual Facebook accounts, they found a post from their programmer and driver, Kevin Wojcik. Entitled "Robots," it had been written at 2:57 that morning.

It's those nights when you wonder what you are doing, and why.

You block out the blinding lights with your hands,
and massage your forehead with your tired palms.
And when your hands smell of dirt, wood, metal, and fiberglass
—it only reminds you
and prevents you from hiding from everything around.
And you think, "I'm just wasting my life right now, this is stupid,"
And you nearly stand up and walk out the door...

But it's on one of those nights that you really think,
and wonder, that you realize what you have gained.
And as you relax in a circle on the floor of the build room,
you see what you have.

Four years ago you didn't know anyone there,
they were strangers, yet now, when you think about where
you will be in a few months,
and ponder the idea that you may never see these people again,
you realize the bonds you have made.
And you look across the room at your creation, where you put
  in hundreds of hours of
your life, and you look back at your friends,
and notice that those hours have done much more...

You realize that they are the reason you are still there.

John Keats, Kevin was not, but if there was a moment when the thirty-one students in the senior class of the Dos Pueblos Engineering Academy understood that they were now a team, reading this note was it.

On March 22, the Sunday before the Sacramento Regional, Chase drove their practice robot around the build room, trying to avoid being pinned by a low, flat, radio-controlled box on wheels. This was the StanBot, named for its creator, who had rigged a chassis with wheels to the controls of a radio-controlled car. The team had transformed the build room into a makeshift mini Lunacy playing field. They had shoved all their tools, parts, and bins underneath the workstations to clear space for the eighteen sheets of regolith they had taped together on the floor. A foot-high perimeter of plywood acted as the walls.

As soon as they had returned to Goleta from their defeat in Long Beach, Amir had asked his students, "Should we continue working?"

"Yes," they said. "We should do everything we can."

Driver practice was key. Chase needed to have an intuitive feel for operating the omnidirectional drive, and the way to gain that was lots of practice. Prior to the Los Angeles Regional, he had only driven the practice robot around the cafeteria. Now he and Kevin were running drills, trying to collect and shoot balls, while the StanBot, driven by other students, pestered them at every turn.

Midway through that Sunday, Amir set eighteen balls on the field, one in the center of each sheet of regolith, and timed how long it took his drivers to collect them. "Go!" he yelled, checking his watch.

Chase began well, gathering a ball in the corner and then spinning away, a maneuver he had had trouble with in Long Beach. Then he started to struggle, missing a few. Amir challenged him: "Where are you going?" Finally, Chase collected all eighteen and Kevin fired them into a trailer. The exercise took close to a minute and a half.

"We need to be able to do this in one minute. Let's do it again."

In the corner, Gabe sat at his laptop, almost finished with rewriting the code to track opposing trailers. The arena lights had thrown off their camera more than he expected. At one point after Long Beach, Gabe and several others had thought they might never get it to work well enough, that all their time programming the turreted shooter would go to waste. But then he just dug in. If they could automatically track trailers on the field, they'd be unbeatable.

Now he thought he had the problem isolated. They had previously determined the distance between their robot and an opposing trailer by measuring the area of a trailer's vision target in the images captured by their camera. The bright lights in the arena washed out some portion of the image, making it appear smaller on camera. The code translated this to mean the vision target was farther away than it actually was. This would cause the shooter to propel balls farther, and they would miss the trailer.

The programmers had had a key realization a few days before. They didn't need to calculate area to measure distance. As an object moved farther away from the camera, its width and height diminished in equal proportion in the image. Therefore, the code only needed to look for the width measurement of the target in the image, a measurement that was rarely susceptible to complete washout from the lights. With this, they should receive an accurate distance calculation between their robot and target every time.

That Sunday, when Gabe finished revising the camera code, Chase and Kevin put on their new glasses. Each pair had an indicator light made from a two-color LED, which was hot glued to the corner of the right lens. A red light told the drivers that their shooter was searching for a target. A green light indicated that the shooter was locked and ready to fire. The new glasses had nothing to do with the new code Gabe had written, but they would allow the drivers to know when they had a lock.

As the drive team took the joystick controls, a pair of teammates pushed the trailer across the regolith. Chase spun the robot until its camera was in view of the trailer's target. The LED turned green. Locked on. When the trailer was shifted left, the turret shifted left. When it moved back, the adjustable hood retracted for a longer shot. Kevin fired, and every ball sank into the trailer.

On Monday afternoon the team met to continue its drive practice and refinement of the vision-targeted turret. They ran a systems check on their practice robot, and they noticed again a strange bug presenting itself. Occasionally when they powered up the robot, a random motor would start to run. Sometimes it was the flywheel, other times a motor on the drive train. They couldn't find the source of the bug, and it seemed to correct itself as soon as all the code on their robot was loaded. They thought it was harmless until the motor that rotated the turret ran uncontrollably.

"We need a protocol for this," Amir told his programmers. They should rig up a switch that would disconnect the turret before they turned on the robot. He warned them of the worst-case scenario, "We'll be in the pit. You won't follow the protocol, and you'll destroy this thing."

His programmers nodded.

As yet another night went late, Amir and his students questioned the sanity of spending this much time in the build room this late in the season. "I feel like we're on that show *Lost,*" Amir said. "They're on this island, then they get off the island, but then they want to go back. It's kind of like what we're doing: We're sick of it—but we still want to be here. It's a perverse situation where we're busting ourselves to win in Sacramento, but winning means we get stuck on the island for longer—because of Atlanta—but it's like we can't *not* try because we love it so much."

As for that sensor connection that had come loose and crippled their robot in the finals, it was now secured with enough hot glue that one would need a sledgehammer to knock it free.

Four days later at the University of California, Davis basketball arena, just outside Sacramento, Chase crawled into their black crate to free the PenguinBot. He loosened the straps holding it down, then, with several teammates, lifted it out. A quick inspection revealed that their robot had survived the voyage from Long Beach with no damage other than a loosened electrical board.

The arena was a modern stone-and-glass structure, much newer and more intimate than the one in Long Beach. There were half as many seats, with easier access to the pits. The floor was a warm gymnasium hardwood rather than the cold, multipurpose concrete at Long Beach. Outside was a college campus with lots of grass and low-sloped hills instead of a corporate park and hotels.

In the pits, Chase rubbed the side of the robot like a father trying to soothe his child, while Turk gave the machine a little pep talk: "It was a tough loss. That foot fault in the first finals match, then a loose sensor. I want you to know it wasn't your fault. I let you down. We let you down. We can do this together."

The team thought they had a good chance to win. On the way to the arena from their hotel that morning, Gabe had read from his iPhone a pregame analysis from an authoritative but anonymous source who posted on ChiefDelphi. "Looking Forward" forecasted the outcomes of this weekend's competitions:

> Sacramento has traditionally been the most defensive of the West Coast Regionals and will likely play similar this year. A group of strong offensive teams will be present though, but they'll likely be gobbled up by the top five or six alliances. Whichever alliance can best survive the defensive specialists and plays the smartest will likely win.
>
> 1717 looks to be the top team headed into the event, and possibly the top shooter in *FIRST*. The D'Penguineers have utilized their turret's potential better than any other team so far (scoring while moving and scoring on robots who are pinning them). 597 and 294 didn't provide enough secondary scoring for 1717 to win in LA, so their destiny may depend on their seeding (or who seeds high enough to select them).

The forecast aside, no D'Penguineer was taking a win for granted that weekend, knowing well from Long Beach that bad luck or the slightest mistake could doom their chance to make it to the Atlanta Championship.

During their first practice match of the day, Chase followed an opposing robot down the side of the field. The shooter locked onto the moving trailer, adjusted itself, and fired a stream of moon rocks at a distance of more than 5 feet. Only one missed.

"I demand perfection," Turk said.

"Well, I feel fuzzy," Gabe answered. His new code had worked.

Afterward on the small field in the pits, Amir had his students run through every system on the robot. Everything was fine.

There were so few problems throughout the second half of the day that Amir was almost frustrated that they had nothing to do. In previous years his team was so busy on practice days that they had to be forced out of the pit before the doors closed. On Thursday, they experienced only two hitches: First, they couldn't get enough time on the practice field to record a new autonomous routine; second, there was some miscommunication between Chase and Amir during matches. For now they couldn't do anything about either.

At 7 P.M., an hour before the arena closed, Amir returned to his team's pit to find several students polishing the PenguinBot with rags and fixing any scuffs on the black anodized metal with a Sharpie. He looked on skeptically.

"Prevents air resistance," Luke explained, his beret perched on the side of his head.

It was a good beginning to the weekend.

An emcee ran onto the field on Friday morning, the first day of qualifiers. He was wearing a massive red ball over his head. "Welcome to this year's *FIRST* Robotics Competition!"

Another emcee ran up to his side. He wore dark reflective sunglasses, a ponytail, and a tie-dyed lab coat.

"Oh, no," Colin said to several of his teammates. "This is going to be rough."

Speeches followed, then the same video that they had seen in Long Beach with Dean and Woodie in close-up, speaking to the crowd like they were all in the same room together.

Outside, Amir sat down against a concrete wall by the loading ramp that led into the arena. His drive team and Gabe sat by his side.

Amir shielded his eyes from the harsh glare of the sun. "We need to communicate better on the field."

"You have to tell me more than 'Go after the red trailer,'" Chase said. "That isn't useful because I've got a limited range of view. I need more specifics. It's confusing."

"Increase your talkativeness," Amir said. "But keep it relevant. If the three of us waste ten seconds talking about things going on in the field that we don't care about, that's thirty wasted man-seconds."

The four made their way back into the arena. The pits were half the size of those in Long Beach but no less chaotic. The tiniest Viking in history, horned hat and all, strode past their robot as they ran through one last systems check. A student dressed in an owl costume walked beside several kids in tall orange hats. Following them was a group in hospital scrubs and a *FIRST* volunteer in a cowboy hat and chaps with a name tag that read "Crazy Nick."

Amir stood by the PenguinBot, studying a scouting sheet that his students had prepared for their first qualifying match.

At the previous regional, the scouts had simply told the drive team what they thought of their alliance partners and competitors. At the Sacramento Regional, they were much more rigorous, both in the amount of data they were compiling on each team and how they would be presenting this information to the drive team.

The scouts were now using a digital camera and portable printer to aid their efforts. For each match, they produced a report with photographs and statistical data on their alliance partners and opponents, including the average number of balls scored by each robot and their human player. This data was based on the practice matches but would change over the course of the day as they assembled more information. They even had little icons for each team that statistics couldn't measure. An icon of a domino represented a "Rube Goldberg machine." The icon with a black-and-green eye represented a team with "crazy eyes." This referred to a drive team that was confused by strategy discussions and might lack a complete grasp of the game.

At 10 A.M., the D'Penguineers were ready for their first qualification match.

A 74–40 victory.

Before their next match a half hour later, Amir identified the one opponent robot to stop and then advised his alliance. "No foot faults and we can win this." Echoing their second match at Long Beach, a robot on their alliance died midway through the qualifier. Their other partner stepped up to play defense. The D'Penguineers made two major scores. Another win: 90–62.

Chase and Kevin returned to the pit. "It's a good start, but we have seven matches to go. We can't get cocky."

"Don't worry, Shaeer," Chase said. "When we're out there, I'm scared out of my mind."

Amir shook his head. "Well, that's not good either."

Then they were off to strategize with their next alliance. On the way they neared a bunch of students and mentors clustered around a flat-screen monitor listing the real-time competition rankings based on win/loss record. Teams checked their rankings after each match. Amir stopped to see that the D'Penguineers were on top. "None of my teams have ever been number one before," Amir said, his chest inflating a little.

Gabe stayed in the pits to work on the shooter's flywheel. In the match before, balls had jammed in the shooter when it had a point-blank shot. When the PenguinBot moved away from the trailer, balls dribbled out. He ran through a systems check but couldn't repeat the problem or find its source.

In the third match, Chase was scatterbrained, driving into corners for no reason and bypassing obvious balls for collection. Kevin missed several easy shots. Meanwhile, the opposing human shooter on CHS Robotics (Team 2437), a short, thin kid with big goggles, who threw the moon rocks overhead like a soccer player executing a throw in from the sidelines, was dead on. Thanks to Turk though, the D'Penguineers still squeaked out a win, 78–68.

They were now 3–0.

In the next match, CHS Robotics and the D'Penguineers were paired together. The opposing alliance was no threat. Another win, 82–44, but the shooter flywheel had jammed again.

The drive team rolled the robot into the pit. With Stan watching, Gabe ran through a systems check, but again they could not reproduce the failure. Then he tested their camera on the practice trailer they had brought on the bus to the UC Davis arena. Everything was fine.

"Are you sure it's close enough?" Amir asked, grabbing the practice trailer by its center post. In his attempt to drag the trailer closer, he accidentally tilted the center post, bringing its vision target almost flush to the camera. Suddenly, the flywheel, which had been spinning forward, stalled. Then spun backward.

Everybody stopped.

"There's your problem," Amir said.

"Yes!" Gabe said. During collisions on the field, the vision target on the opposing trailer was so close to the camera that the code Gabe had rewritten to measure its distance went into territory he had never accounted for. Two lines of code later, the problem was fixed.

Match five was another decisive victory: 90–54. When the PenguinBot came anywhere near an opposing trailer and its automatic tracking system locked onto the target, their team was unstoppable. They were now 5–0.

Before their sixth match, a tough one against two very good teams, Amir brought aside the drivers of their alliance partner, Green Machine (Team 2063). He had read the scouting sheet on the team and watched their previous match. They were rushing— and often missing—to score the seven balls they had stored in their robot at the start of each match. Then they spent so much time trying to pick up additional balls to score with that they were proving to be an ineffective partner.

"Focus on defense," Amir told Green Machine's drivers. "You have two minutes to shoot your seven balls. Be patient. Eventually you'll have a good shot. If you can shut down their best robot and just score those seven balls when the time's right, we have this."

Their drivers executed this strategy perfectly, leaving the PenguinBot free on the field. They won 72–64, the game announcer declaring midmatch that "the PenguinBot's visual tracking system is on the money."

Now 6–0, the D'Penguineers were one of two undefeated teams. Chase was so focused that he didn't even know what match they were in or how close they were to day's end. When the emcee announced the team's final qualifying match of the day, Chase looked up confused. "Already?"

They rolled to another victory, 74–35.

In one of the last matches of the day, the Green Machine shut down the only other undefeated team by repeating the strategy

Amir had taught them. He watched the match with unadulterated joy, both because they had followed his advice and because their match had left the D'Penguineers the lone 7–0 team.

Once the pits closed, the team filed onto the bus. After Amir boarded, Roy, who was back for his second trip with the D'Penguineers, fired up the engine. As they drove out of the parking lot, Amir stood, a hand on the seat in front of him, another holding the microphone.

"Okay, listen up," Amir said, but then Emily yanked on his arm and whispered into his ear. Warned, he began again.

"Okay, my Friday speeches tend to be a low point"—the team laughed, remembering Long Beach, where he told them their fate was already written after the first day of competition—"so I'm going to do this very carefully. I'm going to minimize my insanity by just giving some props."

Everyone cheered.

Amir looked down at the floor, as if to refocus. "I want to say major, major props to the scouting. I'm telling you, we're a different strategizing team because of that work you're doing. It's a whole new world. It enables me as a coach—"

The students broke into a rendition of "It's a Whole New World."

"Oh God, you guys are killing me," Amir said.

"Can we give props to the drivers," a student shouted out from the back of the bus.

"Yes," Amir said.

The team cheered for Chase and Kevin.

"How about Turk?" asked another student.

"Turk! Turk! Turk!" everyone chanted.

"I need to say this," Amir added. "The work that Yidi and Alyssa are doing in the pit with the judges—everyone who's helping in the pit to talk to the judges, that's going really, really well. On every front—the systems-check people, the battery-charging people—everything's great. If I'm forgetting anything, feel free to call it out."

Luke called from the back: "How about Shaeer?!"

"*Woohooo!* Yes!" The team cheered again.

"Thank you. You should know that we are number one right now," Amir said. "We've never been number one before. We've never had an undefeated day."

A festival of clapping rang through the bus.

"And Stan's happy!"

This generated the biggest round of applause.

Back at the hotel after dinner, they were all business again. The scouting captains, Andrew and Stuart, met in their room with Amir, the drive team, Turk, and a handful of others to go over who they should pick the next day during alliance selections. They still had two more qualifying matches, but if they remained undefeated, they would have the key first draft. The scouts ran through the data on the various teams. Amir wanted another big scorer, someone who could add to the forty-plus points the PenguinBot could bring to the match. High on their list for first pick were the Apes of Wrath (Team 668) and the Prototypes (Team 2854). They had a dozen other teams listed for a more defensive second pick, but no decisions were made.

Going into the final day, they *believed* they could win the championship. For the sake of his students and the academy, Amir hoped this would make all the difference.

Sacramento Regional Competition

THE PAVILION, UNIVERSITY OF CALIFORNIA, DAVIS, MARCH 26–28, PART II

It was *so* worth it.

—GABE RIVES-CORBETT

Catastrophic failure! I have to go!"

It was 8:40 A.M. on the final day of the Sacramento Regional, moments before the opening ceremony. Amir passed the cell phone off to a student with Emily still on the line and headed toward the PenguinBot. His jaw had gone slack, and his eyes were wide.

Gabe stood before the robot, his laptop in his hand. He didn't know what to do. During a systems check, the belt that rotated the turret had snapped off. Without that belt, they had no shooter. No shooter, no scoring. No scoring, and their competition was finished.

When Amir reached the robot, he stared at the top plate, not even looking at anybody else in the pit as everybody threw out panicked explanations and solutions for what had just occurred.

"I don't need talking. We need to think. I love you guys, but remember what I said it would be like when we're in the pit, mistakes are made, and the world ends? Well, this is what that looks like."

Amir circled the robot, his head tilted sideways to look inside the hood and under the shooter. He knew what had happened already. During a systems check, Gabe had not disabled the motor that turned the turret. The motor had powered up unexpectedly and rotated way past its limit point. This caused the belt that turned the lazy Susan to tear away from the rivets holding it in place and simultaneously ruined the sensor that gauged the turret's position. Two weeks earlier, Amir had given his programmers explicit instructions to set up a protocol to prevent such an accident and they had ignored him.

"The pits will be closing in fifteen minutes. Everyone needs to begin making their way to the stands for opening ceremonies."

Amir fired off instructions. With so little time, this problem was too complicated for the students to fix on their own. "I need three sheet-metal screws! I need a hand drill! Calipers! Give me a Sharpie! I need a new top belt!"

Daniel Huthsing, who had worked on the shooter team and was now the designated pit boss, searched through toolbox drawers and bins. As soon as he found an item, he handed it to his teammates, who then held it at the ready for Amir. He was the surgeon, they were his assistants.

Needing to replace the belt, Amir drilled out the two rivets still holding the old one onto the turret and yanked it free.

Seven minutes until the opening ceremony.

He wrapped a spare belt around the lazy Susan and squatted down to secure it with three screws. Daniel held a vacuum right below where Amir was drilling to suck in any specks of metal before they fell onto the electrical board and fried their robot.

"Make sure it has the same range of motion," Stan said over Amir's shoulder, watching his every move.

Amir inserted the last screw and then manually turned the turret on the new belt. It rotated without any problems. He also needed to replace the sensor and calibrate everything exactly as it was before, or their shooter would be off aim.

"We're going to be okay," he repeated several times.

A final call came to clear the pits. They still had a lot to do. Amir laced his fingers behind his head and walked toward the ramp outside the arena. He sat against the wall and cupped the lip of his baseball cap to keep the sun from his eyes. Gabe, Chase, and Kevin approached him.

"I'm extremely frustrated right now," Amir said.

Gabe explained that he had a solution to keep the turret from ever doing that again.

"That's fine. You guys are all smart enough to deal with it. But there are a few things I know, and when I say I know them, you have to trust me. And it's always programmers who don't believe me. They always say, 'It'll work out. It'll work out.'"

Gabe lowered his eyes, one hand over his Adam's apple. His throat still burned from when he had almost thrown up on seeing the turret belt rip off their shooter.

Amir noticed that one of the sponsor patches on Kevin's flight suit was missing. He eyed it and asked, "What happened?"

"I was in a panic when you asked for a Sharpie," Kevin said. "When I reached into that pocket to pull one out, I yanked off the patch instead."

"Okay, but it doesn't look so good. I'm obviously not so frustrated I'm asking you about patches," Amir said with a smile, surprising them. "When I go back in, we won't have a lot of time. Here's what I'm going to do." He started listing tools they would need and the steps they would follow to fix the sensor and make sure the turret was calibrated correctly. He was talking to himself as much as he was to the students. He stopped, restarted, changed his sequence, asked Chase and his programmers questions, and listened to their input until he knew how to fix the PenguinBot before the first qualifying match. In that moment, it was clear that the boundaries between teacher and students had broken down. They were equals.

When the pits reopened, Amir and his three students strode back to their robot. They followed the sequence they had set out on the ramp, and within ten minutes they were ready for a test. Kevin sat down in a chair, the operator console in his lap. He moved the joystick, but the turret was still.

Everyone was flabbergasted. The turret should be working now. They were running out of time.

Amir looked at Kevin, and a slight grin crept across his face. "Kevin, you're on the wrong joystick."

Kevin was so overwrought that he had been moving the joystick that controlled the drive train.

They reran the test, and the turret rotated and recentered itself.

"Awesome!" Amir said.

Stan took a close look with a flashlight. "I'm fine with it."

"We need a new battery, and we need to get the scouting team over here," Amir said. "We're ready to go!"

Minutes later, the drive team was rolling the PenguinBot into line for their second-to-last qualifying match of the regional. Amir scanned the scouting sheet while they walked. He met with their alliance partner the Apes of Wrath, a team on the D'Penguineers' short list for the tournament alliance selections. Their lead mentor, Dennis Jenks, a tall, burly man with goatee and a Panama cap, was an aerospace engineer and experienced coach. Their robot, Ape, featured a dumper that could hold more than thirty balls, and their driver was very skilled.

Amir was up front: "I want to see how well our teams play together to see if we'd be a good alliance for the tournament."

"I'd love to see that happen," Dennis said, "but if you need to pick someone else, we're fine with that."

Seconds before their match, a student from Team 2144, who the D'Penguineers were allied with in their next and final qualifier, rushed up to Amir, asking for help with a dead motor.

"Go to our pit. Tell them what you need," Amir said. As the kid ran off, Amir looked toward the ceiling. "When it rains, it pours."

Kevin set up their operator console at the driver station and then stepped back from the line beside Chase and Amir. "I just want the robot to work," he said.

Even though they were facing a weaker alliance and paired with the Apes, the match was tight. In autonomous the opposing human players racked up a double-digit lead. The third robot in the D'Penguineers alliance didn't put up the defense they needed, and Kevin misfired almost a full helix of balls. Only a devastating dump by the Ape onto an opposing trailer clinched the victory: 80–78. Amir rolled his eyes once the score was announced. Too close, but still a win. One match to go to remain undefeated.

When Amir checked in at Team 2144's pit, he found their robot upended while their mentor and several students fiddled with their drive train. Only at the very last second before their match did Team 2144 bring their robot to the field. The D'Penguineers alliance was up against the second-ranked team. If they lost this match, they would each have an 8–1 record. *FIRST* would then rank the number-one team based on tiebreaking rules that factored in which team had endured the tougher qualifying-match schedule.

"Come on, Chase," Turk said. "Let's do this! Let's do it!"

Chase was trying to maintain his calm. "Okay, Turk," he said, staring at the field.

During autonomous, Turk rained down shots on their opponents. Chase drove like he had wanted to since the competition began. Everything seemed to slow down for him. He wasn't looking at the PenguinBot's wheels to know which way to turn. Instead he surveyed the field and maneuvered the robot like it was an extension of himself.

While waiting for the final score after the match, Chase hugged Turk. The real-time score on the projection screen was close, but Chase knew they had the win.

Seventy-six to fifty-two, the announcer declared. Undefeated in qualifications, the D'Penguineers were the top-ranked team going into the elimination tournament. In the stands, the team went wild. Amir walked onto the field. He let go of his breath and clapped, relishing the moment.

On returning to the pit with the PenguinBot, they found their whole team waiting there to congratulate them.

Bryan Heller, their backup shooter, said, "I don't think I've been nine-zero in anything since Boys and Girls Club basketball, and even then—"

"It's a good day to be in a flight suit," Luke said, smiling.

Amir tracked down Stuart, who would soon be on the field to make their team's alliance selections. "Who do you want to pick? I think if the Apes are still in the running, we go with them. I believe you go with coaching experience, and you go with who knows how to win. The Apes know how to win."

Stuart stared at his list of teams while tugging at his long sideburns. "Yeah, they're on our list."

"So I can go tell him? I don't want to string them along."

Stuart thought for a second. "Yeah, let's do it."

A half hour later, Stuart walked onto the field for alliance selections and picked the Apes of Wrath. They accepted with a roar of cheers from the stands.

Stuart had a long wait to choose the third robot in their alliance because of the serpentine draft process. After the eight alliance captains made their first picks in the order of their rank, the picking order reversed in the second round, allowing the eighth-ranked team to complete its three-team alliance before the others. This left the top-ranked D'Penguineers waiting for the rest of the alliances to form before they could make their selection. Being number one was a significant advantage because they had the first pick, but this advantage was tempered by being the last team to make its second pick. This gave some balance to the tournament.

As the other teams made their second-round picks, Stuart feared someone would take his top choice for third robot. The PenguinBot and the Ape were both offensive machines. Therefore, they needed a defensive ally, but one that could at least score a few points each match. CHS Robotics (Team 2473) would fulfill this role. They were the ones with the laserlike human player who threw balls overhead at trailers like a soccer player. Although their robot, Dusty, couldn't score points, their driver had proven effective at hounding opponents.

Finally, the emcee returned to Stuart. The other teams had not seen the value in CHS Robotics, who had finished qualifications with a 5–4 record. Stuart invited them to their alliance.

"I can't believe no one else picked them," Andrew said.

As the stands emptied for lunch before the tournament, there was a slight commotion on the field. Everyone stopped. On bended knee, a team mentor proposed marriage to a volunteer. The crowd applauded as the two embraced.

The D'Penguineers alliance met outside on the delivery ramp and started strategizing. Amir spoke even more intensely than usual. Their whole season rested on these next few matches.

"As long as everyone stays healthy, I think we can win the quarters," Dennis said. "They're just the eighth seed."

"I think we all know how many upsets have been going on this year because of this crazy game. I hear you, but I'm trying not to listen to you." Amir smiled uneasily.

While they waited for their best two-out-of-three quarterfinal, everyone was jazzed up except for Chase. He stood still by the robot, wearing a paper placard around his neck that indicated he was the alliance captain. It looked like a lobster bib from a seafood restaurant. "Seriously, do I have to wear this?" Chase asked Amir.

They continued to strategize until the announcer started listing the teams on each alliance. Amir put his arms around his drivers. "We made it where we need to be. We made it. All the insanity and random pairings are done now. We picked this team."

"Are you nervous?" Chase asked, almost as if he hoped Amir felt as distressed as he did.

"We've got a great team, so I feel calm. Comparatively. We've got to be in the zone. Are you good?"

"I'm ready," Chase said, arms crossed in front of his chest.

Kevin remained quiet beside them, his jaws pulverizing a piece of gum.

Gabe stood outside the barrier separating the drivers from the crowds on the arena floor. He gripped the railings and said to his teammates, "Chase and Kevin have control. Standing here's the worst. At least in the pits, I'm in control."

The buzzer finally sounded. In autonomous, the Ape made it to the corner where their own human player filled their robot's hopper with balls. They were now armed for a big dump. But Turk and CHS's sharpshooter missed shot after shot.

By the start of the driver-operated period, their opponents had an eight-point lead. Their alliance captain, the Robovikes (Team 701), followed this up by dropping several balls into the Ape's trailer for a 34–20 advantage. The PenguinBot returned the favor soon after on an opposing trailer to even out the score.

"Back up over there," Amir said to Chase, urging him to get out of a pin. Their opponents hounded them around the field. "Lots of balls out there. Fifty-six seconds."

Chase maneuvered to collect several balls, then launched three into a trailer. The Ape followed up by dropping a few balls into a moving target to gain a narrow lead for their alliance. CHS's Dusty kept pestering the Robovikes but was unable to hold a pin and found itself a target instead.

"Thirty-two seconds," Amir said. "We're tied right now. Go to work."

An opposing robot circled behind the PenguinBot.

"Get out of there. Commander move. Get out of there!"

Chase escaped just in time. He had a few balls remaining in the helix. He drove straight down the side of the field toward several clustered robots. A second before the final buzzer, Kevin fired two balls into the back of the Robovikes' trailer.

The final score flashed onto the projection screen over the field: 58–54, the D'Penguineers alliance.

"A little too close," Stan said, shaking loose his arms. He was now as carried away in the competition as anybody.

After removing their robots from the field, the D'Penguineers alliance huddled together by the curtain separating the pits from the field. Andrew and Stuart rushed over with their scouting reports.

"We need to know what to do," Amir said, "because that strategy didn't work."

Andrew turned to the Apes of Wrath's drive team. "We saw you guys go for a dump, but most of the balls went onto the floor."

Amir said, "I think we spent too much time trying to jump back and forth between robots instead of scoring on one."

Dennis, the Apes of Wrath coach, added, "If we could get a robot pinned, we can fill that trailer, thirty balls. If not, then it'll be like that last match when we're trying to hit a moving target."

They had to cut the meeting short to circle the pits and get back in line for their next match. Along the way, an eight-year-old boy with team pins covering his hat approached Chase and Kevin. "Are you the first-place team?"

"Yes," Kevin said. He dug into his flight-suit pocket, found a D'Penguineers pin, and handed it to the boy.

"Sweet," the boy said, his eyes widening as if he had seen a superhero. "Wow!"

Before the match, Kevin and Chase rested their foreheads on the top of the PenguinBot. On short glance, one might have thought it was a sign of fatigue. But their intense, almost unblinking eyes revealed how focused they were. All the cheers and buzzers and bombastic announcements from the emcee faded away.

"You've got to relax," Amir told his drivers.

Chase did not look up. "This is *me* being calm."

In the second quarterfinal match, Turk and his CHS counterpart were flawless. All the drivers were on their game. The Ape pinned the Robovikes and unleashed, and the PenguinBot made three short targeted bursts over the course of the match to claim a 76–48 victory.

But there was a big problem: The turret hadn't been working right. Kevin had to operate it manually the entire match, and it wasn't centering itself, so he had to be careful not to rotate it beyond its limit or the belt would shear off again. If that happened, they were done for. They would not have time to replace the belt again as they had earlier that morning.

"We've got to fix this thing right now," Amir said as they carted the PenguinBot off the field before the semifinals. Officials wouldn't allow them back to their pit during the tournament, so they would have to do any repairs while in line for their next match. "We have twelve minutes maximum."

Daniel and several others brought over their toolboxes. Stan and Amir conferred, trying to figure out why the turret wasn't centering. Gabe made it over as well. They had to shout to one another because of the deafening noise from the other matches.

"The pot could be dead," Stan suggested, referring to the sensor on the motor that rotated the turret.

"We need a multimeter," Gabe said. "Maybe a belt jumped?"

A crowd of students circled around the PenguinBot. "Really, do we need seven people here?" Stan asked.

In a rush, they removed the belt leading from the pulley on the turret motor to the lazy Susan. They took off the potentiometer and Gabe tested it. There were no signs that it was dead. While Stan and a student reattached everything, Gabe checked the turret code. "The pot's fine," he said. "The software's fine."

They ran a systems check on the turret, and it was centered. "What's that about?" Amir threw up his hands. They hadn't changed anything, but it was working again. "I love problems that are nonreproducible."

"It works great now," Gabe said, similarly confused.

Amir turned to his scouts. "What's our strategy for the next match?"

Their alliance's drivers and coaches circled one another as Andrew and Stuart handed out their scouting sheets and fed them information. They reported how Dusty continued to have trouble pinning in their previous match. "What should they be doing differently then?" Amir asked.

Dennis spoke to CHS's drivers. "You don't want to chase the robot. You want to go to the point where the robot's going to be. It's like playing defensive back in football: You don't run at the receiver, you go to the spot where he's going." Then he turned to Amir. "And that's the same for you guys too."

The Wildhats (Team 100) led their opposing alliance in the semifinals. Their robot featured a double-barreled shooter but was easily pinned. One of their partners was the Green Machine (Team 2063). Amir had helped coach them in the qualifications to play defense while biding their time to set up the ideal shot of their seven moon rocks. They also had a human player shooting more than 80 percent.

"If Dusty takes out the Wildhats," Dennis said, "then it's two on two. We should outscore them."

Amir also warned the drivers to avoid the center-line position where the Green Machine's human player was positioned.

Finally, they separated to place their robots on the field and line up behind the driver stations.

"What am I doing?" Chase asked. He didn't like the strategy sessions. The torrents of information confused him.

"Drive around, try to collect and score," Amir said. "And don't get scored on. Ever."

Next to them, Kevin kept his arms flat by his side and worried that the turreted shooter might break.

When the emcee, wearing the tie-dyed trench coat, announced the undefeated D'Penguineers, their team jumped up and down in the stands. In the front row, Emily, now visibly pregnant, stood and waved a black-and-white pom-pom. Luke lifted his beret off his head and shook it in the air. Another teammate did several backflips on the floor of the arena, sending the entire place into paroxysms of excitement.

The buzzer sounded.

After an 18–18 tied autonomous, the PenguinBot dispelled fears about its shooter. The automatic tracking system locked onto a target, the turret rotated into position, and Kevin fired, unloading seven balls into the Green Machine's trailer. The score gave their alliance an early lead.

But the PenguinBot was stalked relentlessly by the opposing alliance, allowing Kevin only a couple more shots. Something was wrong with the Ape too. Its hopper was full of balls, but they didn't release when their driver had lined up their robot for an easy dump. Only Turk and CHS's shooter kept their alliance ahead while the Wildhats kept scoring.

The D'Penguineers won 64–60.

"Too close again," Gabe said.

"They're giving me a heart attack," Amir said.

Still, they were a single victory away from the finals.

The alliance gathered beside the field. Dennis came running over. "A ball got jammed. It hasn't happened before, but we're fixing it now. Don't worry."

Amir looked at Chase. "We're going to shut down the Green Machine at the beginning. Pin them."

"Go! Go! Go!" Gabe cheered from the sidelines.

"We can do it," Amir said, yanking at the back of his hair. He told Chase they needed to do everything they could to pin a robot so the Ape could line up for a devastating score.

The emcee boomed, "Who's going to take this match? Red Alliance?" One side of the arena cheered. "Or Blue Alliance?" The other side cheered.

Throughout most of the second semifinal, the two alliances

were evenly scored. With thirty seconds left, it was 54–54. Dennis shot over to Amir. "We've got a ton of balls," he said. "We need a pin."

At that moment, their opponent 766 was near a corner. "Look at 766. That's your kill!" Amir yelled at Chase, hoping to be heard above the cacophony. "Get 766!"

Chase spun the PenguinBot around. He missed the first pin but then rammed his robot front into 766. He kept pushing 766 against the wall as it tried to escape. The seconds ticked down. Ten. Nine. Eight. The Ape came careening down the side of the field. Team 766's trailer was straight ahead.

"Get on them!" Amir yelled. "Get on them!"

"Yes! Yes! Yes!" Chase and Kevin yelled.

With five seconds left, balls spilled out of the Ape into their opponent's trailer. The stands erupted.

Match over: 86–58.

Amir and Chase hurried over to thank the opposing alliances for the match. "It was great meeting you," Amir said to the Green Machine's drive team. "Too bad we couldn't have played together more. We enjoyed working with you."

They headed over to the sidelines to start thinking about their next matches.

"We've come too far not to win it now," Dennis said to Amir and his drivers.

"We're the redeem team," Turk said, joining the huddle before the regional finals.

"Okay, guys," Amir began. "What's the strategy?"

Stuart and Andrew started speaking so fast about their opponent's strengths and weaknesses that they were stumbling on their words.

Amir held up his hand. "But what's our strategy?"

They slowed down and came up with a plan. The Ape would target the TatorBot (Team 2122) and shut it down while collecting balls for one big score. The PenguinBot would roam and score as usual until the Ape had a full hopper of balls. Then Chase needed to set up a pin on an opposing robot so their alliance partner could make a dump. Dusty would make sure not to be scored on while focusing on El Toro, Team 115's robot.

"If there's a super cell in play," Dennis warned, "we have to know and stay away from that shooter."

As the seconds ticked down before the match, the noise in the arena elevated. "We Will Rock You" by Queen blared from loudspeakers, and everyone in the stands pounded their feet.

The alliance drive teams separated and placed their robots on the field. "Zip ties in place?" Amir asked about the battery. Kevin nodded. They could not have a dead robot on the field now.

Amir met with his team one last time. "Turk," he said, "it's all you."

Turk looked up, shocked.

Amir turned to Chase and Kevin. "Whoever the Ape's chasing, we want to help them."

"Buckle your seat belts," Turk said, swinging his arms back and forth, pumping himself up. "Next time we play, it'll be Atlanta."

The emcee introduced the teams. When called, the Apes of Wrath banged on the hard plastic walls separating their driving stations from the field.

Next the emcee said, "Team 1717! The D'Penguineers from Goleta, California!" Amir and his drivers performed their penguin dance, then waved at their teammates in the stands.

"Red Alliance, are you ready?" The emcee pointed at Chase and Kevin before turning to the opposite side. "Blue Alliance, are you ready? Referees, are you ready? Judges, are you ready?" Then he raised his arms and pointed to the packed stands. "You guys up there, are you ready? Let's do this!"

"No foot faults," Amir warned the three drive teams just before the match started.

"Three, two, one, go!"

In autonomous their opponents outshot them by six points, but only because the Ape's human player tossed balls into their hopper instead of shooting. As soon as Chase took the controllers, he spun the PenguinBot around, zoned in on the TatorBot, and made an almost instant fourteen points. Then he pinned the TatorBot in the corner.

"Yes! Yes!" Amir shouted in his ear. "Push them in."

The Ape barreled down on the TatorBot and filled the trailer with so many balls that they spilled out over the sides.

"Whoaa!!" Gabe cheered from the sidelines.

The real-time score slowly ticked up as the officials counted the number of balls dumped, but it was obvious that the D'Penguineers alliance was far ahead. Dusty badgered El Toro, keeping it from any major scores. With a minute remaining on the clock, the Ape and PenguinBot joined forces again for a score, and in the final seconds, Kevin fired another stream of balls into an opposing trailer.

Final score: 112–40.

The whole Dos Pueblos team pumped fists in the air, yelling, dancing, jumping, and, here and there, crying. Saher and Lisa had been screaming for so long that their voices were no louder than hoarse whispers.

"That's the way to do it," Amir said, congratulating his drivers. "One more time."

Daniel, head of the pit crew, called Amir over to see if they needed anything. "I want you to pray," Amir said. "And I'm not even religious."

Only Chase and Kevin were not bouncing around. "Stay calm," Chase said as much to himself as Kevin.

"I'm about to have a heart attack," Kevin replied, his voice drowned out by the Village People's "YMCA."

"Whatever it is you did," Amir told Dennis, "just do it again."

Minutes later, the drive teams were back on the field putting their robots into position and the emcee was revving up the crowd even more. "It's not over until it's over!"

The two alliances exchanged handshakes before the match, then settled behind the line separating them from their drive stations. Amir and Dennis spoke hurriedly before returning to their teams.

Amir put his arms around his drivers. "We'll do this."

"Two alliances here," the emcee said. "Only one of them will be going to Atlanta!"

After the buzzer sounded, the opposing alliance clocked up a quick ten-point lead. Throughout autonomous, Turk was off his mark. Once the drivers stepped forward, El Toro furthered this lead another ten points with an opening salvo.

Chase began to panic. He circled around and found a trailer. "Take the shot!" he yelled at Kevin. "We have a lock."

"There are no balls in the helix!" Kevin shouted back.

For the next minute and a half, the robots seemed to loop around the field, missing shots, collecting balls, and eluding pins. The D'Penguineers alliance closed the lead by a few points, but with thirty seconds remaining they were down 48–32.

Chase circled the middle of the field and then started driving along the wall. The TatorBot pursued him. Amir spotted an opportunity. "You need to help the Ape," he told Chase.

Chase slowed down, seeing what was happening. The Ape, its hopper half loaded, was following just behind the TatorBot.

"Stop! Stop!" Amir said.

Chase halted the PenguinBot in its tracks. The TatorBot rammed into their trailer, while El Toro crashed into their side. The Ape had all the time it needed. It pushed up against the back of the TatorBot's trailer and unloaded twelve moon rocks, for twenty-four points.

"Get out of there! Go!" Amir yelled to Chase, who shifted the joystick diagonally and bulleted out of the pin, no balls scoring on their trailer.

At ten seconds, their opponents had a super cell in play. "Stay away from that side," Amir urged his alliance. The fifteen point super cell missed and rolled away on the field.

Before the final buzzer, the two sides traded a few more shots.

The real-time score on the projection screen had the D'Penguineers alliance losing 54–50—too close to call. *FIRST* officials hurried onto the field to tally the balls in each trailer.

Amir stood off to the side with Dennis. "Let's hope they didn't count your scoring fast enough, like the last match."

On the field, Chase and Kevin forced themselves to change their robot's battery, not wanting another match but needing to be ready if there was one. Turk stood by their side, hands on his hips, eyes glued to the projection screen. Finally, he kneeled down to start loading seven balls into the PenguinBot.

It felt like an eternity passed waiting for the final count.

Then came the announcement: "Scores are in. It's close, ladies and gentlemen, Red Alliance: sixty. Blue Alliance: fifty-eight!"

"Yes! Yes! Yes!" Amir shouted into the rafters. He would have cried, but he was too excited.

Turk turned toward the stands where his teammates were leaping up and down and hugging one another. "Yeah!" he screamed at the top of his lungs before skipping toward Chase and Kevin. Amir and his two drivers had their fists in the air when Turk hopped toward them, his chest out and arms down in an exaggerated flex. An intended chest bump became a bear hug, and in their revelry, Chase and Turk collapsed to the field, bringing Amir down with them. For a moment, they lay sprawled on the regolith, smiling like kings, looking up at the bright lights. "We Are the Champions" echoed throughout the arena.

Amir stood and congratulated the opposing alliance teams and then his own partners.

"Make your reservations for Atlanta," Dennis said. "We'll see you there."

"You don't know how much this means to me," Amir said, almost desperately. "I'm so happy."

Chase and Turk hauled the PenguinBot off the field while Kevin grabbed their operator consoles. By the time they returned to the pit, the rest of the team had already gathered to cheer for them. Gabe pushed through the mêlée. Amir saw him, held out his arm to bring him into a hug. "Was it worth it now?" Amir asked of all the hours, hard work, and perseverance.

"It was *so* worth it," Gabe said, tears falling down his face. He buried his head in Amir's chest.

Stan came over to join in the celebration.

"Was that acceptable?" Amir asked, smiling broadly.

"Yes, that was acceptable," Stan said with a smirk. "Very acceptable."

Chase retreated from the chaos in the pits. He made his way through the arena's double doors leading to the delivery ramp. He stripped off the top half of his flight suit, his white T-shirt underneath soaked in sweat, and then sat down in the sun. He closed his eyes. The victory had still not sunk in completely.

He was happy about what the team had accomplished. They

had built such an incredible robot, Chase thought, it deserved to go to Atlanta. He almost felt like he was only there to show the world what the PenguinBot could do.

When Kevin joined him out on the ramp, Chase said, "I could drive six more months and not live up to the potential of our robot. It's that good."

Everyone assembled on the bus for the seven-hour ride back to Goleta, their championship medals hung around their necks.

"We're on our way to Atlanta!" Amir said, standing at the front of the bus with the microphone.

The team clapped and pounded the bus floor with their feet.

"I don't have much left in me, but I have to tell you that I'm excited. I'm proud of all of you. I hope you're proud of what we've achieved."

In the back of the bus, one student asked, "Can we wear our medals to school on Monday?"

"I think it would be cool if you do when you wear your flight suit before we go to Atlanta." Amir smiled. "Wearing medals on Monday may be a little dorky."

"You mean like my hat?" Gabe asked, donning his orbit-ball top hat with its chin strap followed by the lobster bib that Chase had worn throughout the tournament.

"You look like something out of Dr. Seuss," Amir said. The team broke into laughter, then he settled them down again. "When I fell on the field after the final match, it was out of pure joy."

"Awwwwwwww," several students responded.

"Scouting was awesome. The pit crew was awesome. Everything was awesome."

The team cheered and applauded one another.

"It was a perfect tournament, you guys. Fifteen and zero! Woohoo!"

FIRST Championship
GEORGIA DOME, ATLANTA, APRIL 16-18, PART 1

This is it. We're here, and we're never going to be
here again.

—CHASE BUCHANAN

ragedy struck the engineering academy soon after the D'Penguineers returned from Sacramento.

On April 1, Lindsay Rose, a freshman in the academy, had a surfing accident at a local beach in Goleta during spring break. Fellow
surfers found her floating facedown in the water and dragged her
out. Paramedics rushed to the scene and managed to revive her.
She was hurried to a nearby hospital and put into a drug-induced
coma, but she died four days later, on a Sunday morning. Amir
learned of her death that evening at an academy board meeting.

He knew Lindsay mostly from what he saw of her in class. His
Introduction to Physics class was not easy for her. She was earnest
and worked hard to understand. The day before he heard the news
of her death, Amir had been grading her most recent test. The
score raised her class grade from a B+ to an A–. Amir had thought,
She's really getting it now. He had been looking forward to sharing
this with Lindsay when school resumed after the break.

That Monday, April 6, he had to face her classmates in the first period of the day. Counselors came to the room.

"We've some very bad news to tell you," Amir started, his voice cracking. He tried to continue but couldn't. Tears fell from his eyes. The counselors stepped forward to read a statement about her death, while Amir retreated behind the whiteboard.

He got himself together enough to ask the counselors to leave. He slowly closed the door before turning to his students. He was their teacher. He needed to lead his students through this situation, but at that moment he felt woefully unprepared.

At last he wiped the tears from his eyes and forced out the words. "I want to help you through this. I'm not sure how. But we're going to grieve together. Take it one day at a time, and somehow we'll get through this. Together."

When the period bell sounded, the students filed out. One of them left a rose on Lindsay's desk along with a note that read, "We'll miss you." By the time Amir saw the note, his robotics team was already coming in for their second-period class. The rose and note sent Amir into another spell of crying, this time in the middle of the room. He explained through his tears, "One of my students died."

His seniors were struck silent, both by the news and Amir's state. They understood how much their teacher cared about the 128 students in his academy, but seeing him so affected drove it home for them. The team knew Lindsay but only a little. She had been one of the underclassmen who came down to Long Beach to cheer for them. But Amir saw her every day, had recruited her to his program. Knowing how difficult this must have been for their teacher, his seniors treated him gingerly as they prepared for Atlanta.

The two weeks between the Sacramento Regional and the

FIRST Championship were an emotional roller coaster. Amir and the team had come back to Dos Pueblos flush with victory. The death of Lindsay sobered the students and sank their teacher into despair.

Many who knew Lindsay, and even those who had only heard of her tragedy, wanted to send flowers or a small gift to her family, some expression of sympathy. Her parents asked that people instead send a donation in their daughter's name to the Dos Pueblos Engineering Academy. They wanted future students to have the same opportunity Lindsay had to thrive in the environment of learning Amir had created there.

Amir was surprised and moved by the gesture. Before the team left for Atlanta, he asked Lindsay's parents for one of the rose stickers that her softball team had made after her death in order to place it on the PenguinBot. Amir wanted to win in Atlanta for her.

It was 6 A.M., Thursday, April 16, practice day at the *FIRST* Championship. Chase lay in bed in his room at the Super 8 motel in downtown Atlanta. Kevin was already showering in the bathroom, but Chase stayed under the flower-patterned comforter. In West Coast time it was the middle of the night, and he was exhausted.

The journey from Goleta the day before had been long. Chase had packed in his suitcases his share of the hundreds of pounds of tools, spare parts, and other items the team needed to bring in their checked luggage. The two flights to Atlanta took forever. Then everything had to be unpacked and sorted, and several of them needed to make a preliminary visit to the pits at the World Congress Center next to the Georgia Dome. Afterward, the team wandered downtown Atlanta for an hour looking for a place big enough to eat together. When they returned to the motel, Chase wanted to crash, but everyone was so hyped up that it was hours before he fell into a sleep made restless by dreams of driving the PenguinBot into walls.

Now he was staring at the popcorn-textured ceiling of his ramshackle motel room. Cigarette burns dotted the fabric of the chairs. The carpet was dust brown. The beds were lumpy. The drapes came straight from the seventies, and the air smelled musty. And he soon learned that the walls were paper thin as well.

"Let's go, baby! Let's go, baby!" he heard through the ceiling.

It was then that he realized he had been sleeping below Turk. Chase sighed, crawled out of bed, and donned his black flight suit.

The buzz downstairs perked him up. There were several other teams staying at the motel because of its proximity to the Georgia Dome, and they milled about the mezzanine floor eating breakfast and talking about the competition to come. His teammates were huddled around a few tables, discussing a post on ChiefDelphi that described the D'Penguineers as one of four teams in their division that were "locks" to advance far in the championship.

At seven-thirty the team met Amir in the lobby. He had been feeling under the weather the night before but was now just plain sick. His eyes were glassy and his voice was uneven. "I can't talk much. So when I do, you need to listen. We're going to be carrying our stuff to the site. If you see someone struggling, help them out. We're in this together, and we're going to work together. Remember"—he grinned—"we're a lock. Let's live up to it."

Everyone grabbed a box or crate or stray banner, and like a pack of mules, they started their twenty-minute journey to the Georgia Dome. It was a chilly, but clear morning. One student struggled to get his arms around an awkwardly shaped box containing a printer for the scouts, and others, like the slight Angie Dai, strained to carry what amounted to a fourth of their weight.

Gabe sidled up to Chase as they crossed the street. "Do you want to hit people?" he asked.

Chase didn't know what he was talking about, but Gabe had one of his mischievous looks on his face.

"Shaeer said that if anyone was goofing off on the way, I've got permission to hit them." Gabe laughed. "Even cooler, I can bestow this same privilege to anyone I want."

There was no need for physical blows. Amir led the way through Centennial Olympic Park, his students hurrying behind him, working under the theory that the faster they got to the stadium, the less time they'd have to strain under their burden. They were excited too. Streams of students from around the country and the world flowed toward the Georgia Dome in the distance. They eventually veered off to the right through a bank of glass doors into the World Congress Center. Several city blocks wide and long, the center was a behemoth. The team crossed a long passageway, rode down a couple of escalators, turned a corner into an even longer passageway, and walked down another until they eventually reached the pits.

There was a huge map out front to help teams locate their individual pit among the 348 team pits at the competition. Amir hurried past without looking, since he had uncrated the Penguin-Bot there the day before. The hall was so massive that from its center point one could barely see either end. Separating the pits were four different-colored curtains that identified the divisions of teams: Galileo, Curie, Archimedes, and Newton. Projection screens, which would soon list their qualification matches and team rankings, hung from the ceiling.

In their pit, the team dropped their equipment with sighs of relief. Then the thirty-one students, Amir, and mentor Danny Lang (Stan had chosen to stay home because of his work and family), crowded around their robot in the 10-by-10-foot space. Everyone had donned safety glasses, as required. Colin and Chase had bought a thick-rimmed style for the occasion, making them look like young Buddy Hollys.

"Means of production," Amir said, waving off those students on the team who were inessential to organizing their pit or preparing the PenguinBot for competition. For the pit, they had to hang banners; assemble shelves for the tools and parts; and build a cart out of two-by-fours to transport the robot to and from the Georgia Dome. For the robot, they needed to run systems checks and install an aluminum bar over the intake roller. The collisions in Davis had bent the chassis frame, and they wanted to protect their ball collector from future blows.

Chase and Luke ventured off, eager to check out the scene. They were amazed to find two regulation-sized practice fields, nothing like the little patches of regolith that they had become accustomed to at the regionals. They wound their way around the long rows of pits, stopping here and there.

One of the first teams they came across was 2Train Robotics (Team 395) from New York. Their robot looked like a half-broken hodgepodge of parts, but the team had made it to Atlanta after an against-the-odds win at the Philadelphia Regional. Their driver, Gabe Ruiz, looked very intense. He had cut his hair into a Mohawk and was struggling to straighten the bent aluminum frame of their robot with his bare hands. Chase and Luke wished them luck.

Mostly, the two D'Penguineers checked out previous champion teams that they knew only by reputation. In their lime-green T-shirts, the ThunderChickens (Team 217) from Detroit had a machine with a turret and helix that looked similar to the PenguinBot. This was the team whose mentor Paul Copioli built robots that built robots. Team Hammond (Team 71) had a fierce crab-drive train on their famous Beast. The Simbotics (Team 1114) had a huge bin that was able to treat their balls as a fluid that delivered straight into a shooter. It was the design Amir had pushed his team to go for at the beginning of the season, but that they had never succeeded in building.

"Looks like your turret can go all the way around?" Luke asked one of the Simbotics' mentors.

"In case we have to shoot on ourselves." He winked.

As they moved from pit to pit, Chase became more fired up with each stop. "After seeing these robots on the Internet, and now seeing them, for real," he said to Luke, "you realize *this is it*. We're here, and we're never going to be here again."

An hour later they returned to their pit. The contrast between them and the other teams they had seen was pretty stark. Many of them had numerous mentors overseeing the robot, while Danny was the only one in the D'Penguineers pit, and he was mostly sidelined by their teammates who knew as much or more about the robot as he did. Amir would be busy throughout the competition as the coach of the drive team. The students knew that they were responsible for the robot.

With a clipboard under his arm, Stuart came into the pit soon after Chase and Luke arrived. He and the other scouts had snapped photos and interrogated most of the eighty-seven teams in their division.

"It's a lot of teams," Stuart said, pulling at the ends of his goatee, which had begun to fill out. "We're looking at a pool that's twice as big as Sacramento was. There are four of these pools."

"Who looked good?" Gabe asked.

Stuart checked his clipboard. "God, there are so many teams."

From the terrified look on his face, it was clear that they were all good.

Turk tried to lighten the mood. "I kind of feel like we're under the radar here. A lock? Who even said that? It's a compliment, sure. But I have my own lock; it's a lock that the first day is going to be boring. That's a *lock*."

Turk was off mark on his prediction for the day—way off mark.

The long march pushing the PenguinBot from the pits to their first practice match in the Georgia Dome built up the excitement. On the quarter-mile walk down a cold concrete path split into two lanes, for incoming and outgoing traffic, they passed teams returning from the stadium. Each band of students and mentors rehashed what went right, what went wrong, and what they could have done better. After walking along stretches of loading docks and traveling down and then up long sloping ramps, the D'Penguineers reached the tunnel that led into the stadium.

"Wow!" Kevin said, his neck craned as he looked up at the twenty-seven story dome covered in a white translucent fabric that acted like a giant skylight.

"I can't wait to enter this place tomorrow and hear everybody go nuts," Turk said, taking in the stadium's seventy-thousand-plus seats.

Chase was speechless. He turned around and around, the enormity of this championship hitting him. The Long Beach Arena had seemed big to him. The Georgia Dome was seven times its size, and the site of events as major as the Super Bowl and the NCAA Final Four.

The stadium floor was sectioned off into four Lunacy fields, each with its own projection screen, scoring table, and an array of cameras and lights. Huge curtains hanging from wires separated the fields from one another. The best teams in *FIRST* had come to compete on these fields. The winning alliance from each division would then meet on Einstein, a field to be constructed in the center of the stadium for Saturday's finale, to determine the world champion. Tens of thousands would be watching. Television-news crews and scores of journalists would be on hand for the event.

Gabe took his first step inside the Georgia Dome in stride. He had competed there several times in the tae kwon do national championships. Now wearing a flight suit instead of a *dobak,* he snapped photographs so that he could calibrate the robot's camera to the lighting and colors in the dome.

"I get cheery whenever I come out here," Amir said before prepping his drivers for their first scrimmage. He wanted them to collect two empty cells from the center-line human shooter and deliver them to the corner for the super-cell exchange.

In their previous competitions, the D'Penguineers had not attempted to use the fifteen-point super cells. Before Atlanta though, they had watched several regional competitions online and seen how well some teams were using them to their advantage. At the *FIRST* Championship, Amir was sure the super cells would be an essential strategy for teams.

Chase and Kevin had run through so many driving drills in the past two weeks that they had driven the StanBot into submission. It had died the night before they left Goleta. As they set up for their first match, they were more confident than they had ever been in their piloting skills.

But during the practice match, when Chase tried to collect the two empty cells from the field, he had trouble positioning the robot and left both balls on the field. Still he and Kevin scored on a couple of trailers, keeping it from being a total loss. Scouts from other teams were watching every scrimmage, judging who they might want for alliance selections, so it was important to look good even in these matches.

Afterward, they rushed the robot from the field and back the quarter-mile to the pits to work on their autonomous mode. In their regionals, they had recorded a single routine: drive from the corner, spin 180 degrees, and stop. Now competing against the very best teams, they needed to be able to do different maneuvers from any of the six positions their robot might start from on the field. Every second of a match would be essential. Starting positions and maneuvers in autonomous would be as critical as the opening moves of a chess game.

After Sacramento, Gabe had refined their autonomous code, making it easier to record and play back individual routines. His code now allowed them to assign individual routines to a ten-position dial on the robot. Their makeshift field in the build room hadn't been big enough to do any of these recordings. Now they had a limited amount of time on the pit's regulation-sized practice field to record them. The first one would have the robot drive from the corner, spin 180 degrees, and oscillate back and forth to throw human shooters off their mark.

"Chase, I'm going to say 'Three, two, one' from the field," Gabe began, speaking beside him at the driver station behind the clear Lexan wall. "You'll hit the button to record, drive for fifteen seconds, then we'll set up for the next routine."

"What am I doing after that?" Chase asked, looking skeptical.

"I don't know," Gabe said.

"Next," Amir explained, "you're going to start in the center-line position and come straight over into the corner, where you will unload a super cell to Turk."

"Okay, turn the robot on," Gabe said.

"This first one, you're going to come out of the corner," Amir said, but Chase stared at him like he was speaking gibberish.

In the midst of this confusion, Adam Freeman, the drive coach for HOT (Team 67), was watching. The twenty-nine-year-old GM engineer looked every bit a former college jock, and he always wore the same outfit at competitions, for good luck: black Adidas soccer shoes with three white stripes, a pair of Levi's with a hole in the knee, and a black HOT hoodie.

The last-minute changes his team had made to their robot before their first competition paid off throughout the season. The GM-sponsored team had claimed the Michigan State Championship, and they were one of the other "locks" in the Galileo division. Adam was looking to see if the D'Penguineers might make a good future alliance partner.

At last, Gabe and Chase were ready, and the controller started recording the routine when the PenguinBot drove out of the corner. But just as their robot spun around, one of the two other robots on the practice field crashed into it. They had to begin again.

Ten minutes later, they finally finished one routine.

"Three, two, one, go," Chase said at the start of the second. He pushed his joystick left, maneuvering the PenguinBot from the middle position to his own corner. Amir and Gabe were giving him instructions simultaneously; Chase was asking questions, and he ended up slamming the robot into the wall.

"I think we're going to have to rerecord that one," Chase said.

After the rough start, they settled into a rhythm. Amir ran around the edge of the field, directing where the robot should go. Gabe made sure the controller was recording and timed the fifteen-second autonomous routines. Chase drove while being assaulted by conflicting instructions. They recorded routines to move the PenguinBot from the corner to the center line to collect an empty cell from their human player. They recorded evasive and block maneuvers. There were so many that Gabe and Amir began to confuse the number of the routine on the ten-position dial and its actions.

"The ninth routine is fine. Then what did we do?" Amir asked.

"What I said was eight is actually six, but nine is—" Gabe

stopped, realizing the madness. "Is this what an elite team looks like?"

They were out of time on the field.

News in the pits of an impending visit from Dean Kamen pushed the autonomous drama out of mind. A *Popular Mechanics* photographer came by wanting to feature their team and the PenguinBot in a shot with the *FIRST* founder. There was a stir of voices as Dean approached in his trademark blue jeans and denim shirt. Students and mentors vying for an autograph surrounded him. Dean looked happy as could be to meet any and everybody.

"You've changed my life," one student said as Dean signed the back of her pink team jersey.

A mother said much the same of her son.

"How many years have you played?" Dean asked many of the kids before following up with "You're coming back as a mentor, right?"

Chase, Turk, and several other D'Penguineers jockeyed into position as the photographer staged Dean to the side of the PenguinBot while moon rocks flew into the frame of the shot.

When they were finished, Turk ran back to the pit for one of the team's balls, while Kevin retrieved his driver console. A Dean Kamen autograph was a prized possession, and he signed everything from mascot costumes to goggles to hats. At one competition, he penned his name on a student's leg, and the teenager beelined to a tattoo parlor to have it etched into his skin.

After having pushed his way to Dean, Turk came running back to the pit, palming the moon rock like a basketball. "Right there, baby. *Bam!* Dean Kamen on my moon rock."

Chase was less impressed with Kamen than he was with being in a *Popular Mechanics* photograph. He had been subscribing to the magazine since he was eleven years old.

Amir broke their reverie. They were fifteen minutes out from the second practice match. They rushed their robot back to the Georgia Dome. It was clear now that whoever designed the wide concrete pathway connecting the two structures had motorized transportation in mind. The quarter-mile march pushing their robot was already becoming tiresome.

From the stands, the team could not understand what was happening in the second match. When Chase took control of the PenguinBot, he drove straight past opposing trailers. He wandered around the field, missing balls ripe for collection. He glided straight into pins. It was a disaster.

Amir and Kevin thought Chase had been struck senseless. "I'm going to stop and shoot on this robot," Chase said at one point, only to barrel past it, way off target. "I'm coming around," Chase said a few seconds later before smacking into an alliance robot.

Amir and Kevin exchanged puzzled looks. It was like watching a movie with the wrong soundtrack playing. Nothing was happening when it should.

"What was that?" Amir asked, not even directing the question at Chase because he couldn't believe his driver could be so confused. "What's going on?"

Their scouting captain, Andrew, hurried down to the field, his face flushed. "Chase, you can't be messing around. Other teams are scouting us. We have to look good."

Chase showed only modest improvement in the next match.

"Do you think you could have done better than that?" Amir asked his driver.

"I'm not sure why that happened," Chase said. He was embarrassed and mad at himself. *Maybe it was nerves,* he thought.

Everybody on the team was bewildered as well. This was not their driver. Occasionally at the regionals, he had veered off in a direction he was not supposed to, but nothing like this. If they went into qualifications the next day driving the same way, they would not be able to compete.

Amir and the drive team hurried back to the pits for their second scheduled slot on the practice field to work on autonomous. During the walk, Amir wracked his mind trying to figure out how to help Chase do better. But he was stumped.

Autonomous maneuvers descended into mayhem as well, primarily because Amir was still trying to understand why Chase was having trouble.

Then, on the practice field, as Gabe and Chase tried to communicate through the driver-station wall about where the robot should go, Amir had a revelation. Chase was talking while he was trying to drive, just as he had been in the practice match. Talking continuously during matches was a new strategy that they had decided to implement in Atlanta to improve their communication. Instead it had caused a complete breakdown.

Amir now also understood those moments in Long Beach and Sacramento when Chase had driven off course; he had probably been speaking then as well.

Once they finished recording the autonomous routines, Amir brought Chase and Kevin to a secluded part of the World Congress Center. "You know when I'm working on the robot in the pit, and someone tries to talk to me, and I'm like, 'Shut up, shut up, shut up'? It's because I need to focus. My mind is in a quiet place. I can't deal with talking and focusing like that. Chase, you're the same but for different reasons, because of your learning disability. You're not able to talk and drive the robot. Talking takes up all your focus. When you do, everything else falls apart."

Chase looked off in the distance, like he was rewinding through every match they had planned since the season began. "That makes so much sense," he finally said.

"Your mind won't allow you to talk and drive at the same time.

You need to focus one hundred percent on one or the other. So no more talking."

"Okay."

The three returned to the pit.

At day's end, the D'Penguineers ate dinner together at the CNN Center food court while sifting through the scouting data on whom they were paired with and against. They stumbled back to the Super 8, many with sore feet from all the back-and-forth between the pits and the stadium. They met on the motel's mezzanine, where Amir gave them a few words of encouragement.

After the long day, they were all tired and needed the pep talk. Andrew then said, "We've looked at our schedule, and none of our matches are easy. But we're not against HOT or Team Hammond in qualifications, so you should be thankful for that."

"There are a lot of good robots out there," Amir said. "This is not like a regional. It's crazy tough out there. So, seriously, we really, all of us, need to get to bed. The drivers are going to walk about eight to ten miles tomorrow, so we've got to rest up."

The team applauded, unsure what else to do. Tomorrow they would face some of the best teams in the country, and they were more than a little scared.

Amir smiled. "Yeah, that's right, clap for walking miles."

On the first day of qualifications, the pits were alive. Throughout the World Congress Center, alliance teams huddled together, hashing out their strategy before matches. Calls over the public-address system for teams needing spare parts or tools became more urgent. *FIRST* officials zipped down the long rows atop Segways. Students and mentors circled their robots in full team regalia: Merlin hats, yellow bees with smiley faces, nun's habits, chicken suits, cowboy outfits, fuchsia-colored alien costumes, and T-shirts of every imaginable color.

Everywhere there were herds of kids from fourth grade and above who had traveled to Atlanta from around the United States and scores of other countries. They had come to compete in the championships for the *FIRST* LEGO League and the *FIRST* Tech Challenge, the junior siblings to *FIRST* Robotics. With their friends, they spoke wistfully of one day competing in the big leagues, maybe even on the Einstein field.

Everybody in the pits seemed in a constant state of motion, whether marching to or from the Georgia Dome, hopping up and down in excitement, running on some almost forgotten errand, fixing a broken robot, or stealing a moment for breakfast or a quick call home.

In their black flight suits, the D'Penguineers looked all business. While inspecting their robot one last time before their first match, Amir was surprised by a visit from Adam Freeman, the HOT coach. They had never met before. After brief introductions, they got straight down to competition talk.

"We've a couple of alliances in qualifications that are so good we wouldn't have a chance of getting them if we picked them ourselves," Adam said.

Amir admitted that his team's pairings looked far more difficult, but so goes the random computer algorithm that selects alliances.

"When our scouts measure teams," Adam said, "they don't look at win-loss records. They look at how the human players do and how the robot performs."

"That's how we do it," Amir said, encouraged by the mere fact that HOT had come to talk to his team. Andrew and Stuart had already determined that the Michigan team was the best in their Galileo division. If the D'Penguineers were allied with them in the division tournament, they would surely reach Einstein.

"I saw you guys during your practice session yesterday recording autonomous. It looked like you knew what you were doing."

"We wanted to set up routines to collect the empty cell, but I think we're going to stick with what we're good at."

They shook hands, the mutual wooing over.

Then Amir went straight into the strategy session with the alliance for their first match. They were up against very tough opponents. "Here are the numbers," Amir said, showing his scouting sheet to his partners. One of the drive teams was so nervous about being late to the match that they had brought their robot to the talk. "Based on yesterday's practice session, our scouts calculated the average scoring ability of our alliance's teams versus our opponents. As you can see, if we try to outscore them, we're going to lose. So we're going to turn this into a defensive match."

After Amir took measure of who in their alliance was best at pinning, he set their robot-to-robot defense. They also decided that their alliance would run an empty cell to Turk for a super-cell exchange and potential fifteen-point score in the last twenty seconds of the match.

"The first time I get the super cell is in Atlanta," Turk said afterward to Chase and Kevin, punctuating it with a sarcastic two thumbs-up. "Thanks, guys! No pressure here."

They returned to their pit to find the PenguinBot on blocks with Gabe on his hands and knees in front of it. While running each autonomous routine, he watched the movement of the wheel modules. One of them had been recorded incorrectly. Amir said they wouldn't need the routine, but the glitch unsettled everybody.

The walk to the stadium did little to settle Amir's unease. After the disasters the day before, he and Kevin were still worried about how Chase would perform. They were not alone. Chase thought again and again about everything the team had done to reach Atlanta. At this critical hour he didn't want to embarrass the PenguinBot, his teammates, Amir, or the academy by falling apart now. He was a mess.

They entered the stadium to a cacophony of noise. With the four division fields came four sections of stands cheering for their alliances, four emcees revving up the crowds, four sets of buzzers and horns starting and stopping matches, and four lines of drive teams and their robots awaiting their matches.

Chase kept one hand gripped on the PenguinBot's frame, looking like he might collapse if not for its support. His legs shook.

Time seemed to accelerate, and they were already placing their robot onto the field. The white rose sticker for Lindsay was on the Lexan outer cylinder of the PenguinBot.

Everyone took positions for the match. Amir stood behind Chase and Kevin. There was no talk. Before the buzzer sounded, Turk touched the bin of balls by his side. A referee informed him that he had just earned his alliance a ten-point penalty. Instantly pale and unable to speak, he looked toward Amir, who didn't understand what was happening but knew it couldn't be good.

Their new autonomous routine to oscillate the robot worked as planned, throwing off their opponent's human player. Their robot-to-robot defense also proved successful. Showing none of his nerves, Chase swerved around the field for several scores while evading any shots on their trailer. Their shooter's automatic tracking system was flawless as well. Their alliance held a lead throughout the match and won 72–62, despite the ten-point penalty and a super-cell shot missed by Turk.

As they rolled the robot out of the stadium, Chase started to move faster and faster. He clutched his stomach, and his face drained of color.

"Oh my God. I'm going to throw up," he said, rushing to the nearest trash can.

Gabe grabbed the frame of the robot to keep it from barreling out of control down the ramp. Amir hurried over to Chase.

The spell passed.

"I thought my legs were going to give out because they were shaking so bad," Chase said.

"Mine were too," Amir said, then looked to Kevin. "Were yours shaking?"

"Oh yeah," Kevin said.

"Me too," Turk said.

They all looked toward Gabe.

"Yep."

But the D'Penguineers had survived the first match. Now it was time to show everyone in *FIRST* what their PenguinBot could do.

FIRST Championship

GEORGIA DOME, ATLANTA, APRIL 16–18, PART II

This means everything.

—AMIR ABO-SHAEER

When the D'Penguineers drive team returned to the pits from their first match, they had only minutes to strategize with their next alliance. One of their partners was Raider Robotix, a veteran *FIRST* team. Their mentor wore a snug backpack, a baseball cap, and dark sunglasses. He carried himself with a little bravado. When Amir asked if their robot was pinnable, the mentor said, "Not happenin'." They decided to have their other partner focus on the opposing alliance's best scoring robot, leaving the PenguinBot and the Raider Robotix's Evil Machine on offense.

Then they were marching back to the stadium again. Moments before the match, Amir asked Chase, "How do you feel?"

"I feel like if everyone does their part, we'll be okay."

"No, I was asking how *you* were feeling."

"I'm not thinking about *me* right now. I'm focusing on what we have to do."

Their alliance rolled to a 94–56 victory.

After the match, Chase placed his hands on Kevin's shoulders, bowed his head, and sighed. They were all relieved that their new communication strategy was working. Turk had redeemed himself, lofting one shot after another into trailers until their opponents realized the destruction he was wreaking on them and avoided his corner of the field. The D'Penguineers were 2–0 and, at least for now, ranked second in Galileo. Amir was clear about their chances in their third match. "If we win, we are *gods*."

While most teams broke for lunch on the patches of grass between the World Congress Center and the Georgia Dome, Amir had his drivers and programmers on the practice field in the pits, refining their autonomous yet again. It was risky changing these routines in the heat of competition. They would have no time for thorough testing. During the first qualifier, Gabe hadn't been able to look at the field during autonomous because he was not sure the new routine would work. He promised to *walk* home to Goleta if it hadn't.

But now there was a potentially game-changing autonomous routine they wanted to record. The robot would move to an alliance corner. There, a human player would roll moon rocks into their robot through the low opening in the field's wall, filling up their helix. With twenty-four balls ready to fire, they could overwhelm their opponents in the first few seconds after Chase and Kevin took the controls.

On the practice field, Amir ran to and fro as he realigned the robot again and again in its initial position until Chase had driven the PenguinBot exactly as they wanted for the autonomous routine. Gabe and the drivers then raced up the stairs to inhale sandwiches before the competition recommenced.

According to the scouting reports, they had no chance of winning their third qualifier. They were outgunned, and neither of their partners had shown much taste for defense in previous matches. Amir gave his now almost rote speech to his alliance drive teams about how they should focus on defense throughout the two minutes, only shooting the seven balls their robot started with when they had a trailer lined up.

On their hike to the stadium, Andrew called Chase, who passed the phone to Amir.

"Andrew...Andrew...," Amir said, unable to make sense of what his scout was saying because he was in the stands and speaking too quickly. "Be quiet...listen...shut up...I can't hear you."

Andrew found a quiet place in the stadium and calmed down. Amir then relayed the new information to his alliance.

"My scouts told me the only way these three robots can score is if our trailers are totally stationary. So keep moving. They also say we should let one of you guys run two empty cells to our human player."

Within seconds of the match's start, one of their alliance robots died. "Team 1195 isn't moving. It's not looking good," the emcee said. "They're a sitting duck. They're a sitting duck." Halfway through the match, their other alliance robot died too, right in front of an opposing human player. Chase and Kevin performed brilliantly, but the loss was inevitable: 117–68.

Still, Amir and his drivers were upbeat, since they had scored almost all of the points for their alliance. "That was impressive," Amir said. "I wouldn't want to be against us."

Their scouts shared that same feeling, believing other teams judging potential alliance partners would not fault them for their loss. They also joked that the camera crews must love them because the PenguinBot was featured time and time again on the projection screen during their matches.

"It's funny," Colin said, always handy with a pointed remark, "how looking good is almost as important as doing good here."

In their fourth match, they were paired with Team Hammond (another "lock"), the only four-time *FIRST* champion. Chase was excited to be allied with such a storied team, particularly since their Beast was an offensive dynamo. But he soon lost his enthusiasm.

Their coach, a tall, big-boned giant with green foam plugs sticking halfway out of his ears, had a plan, but it had nothing to do with working together to win the match. If their alliance was way ahead, he wanted the PenguinBot to sit still on the field while the Beast shot into their trailer to run up their opponent's score (but not enough to lose the match). If there was a tie in qualification rankings between two teams, *FIRST* gave the higher ranking to the team whose opponents scored more points.

Amir was absent when this plan was first raised, but when he learned of it, he was clear to the Hammond coach. "You're Team 71," he said. "Everybody knows you. No matter how you do in qualifications, you're fine. We're Team 1717. Nobody knows us here. We can't sit for an entire match. We need to demonstrate what our robot can do. I'm fine to help out with your ranking score, but I want to make sure we win this match first."

The coach refused to be swayed from his strategy, and the two had a polite but strained exchange that ended with Team Hammond's mentor barking, "Enough said."

The match was total chaos, and the D'Penguineers escaped with a narrow 93–82 win.

In the spare few minutes before their fifth and final match of the day, the team returned to the practice field to work on their autonomous. They brought over the human player from one of their alliance partners to show him how to load the PenguinBot from the corner position. The short, thin kid with glasses and too much hair rocked from side to side, he was so excited about participating in this strategy. His team had earned a spot in Atlanta after winning the Rookie All-Star Award at their regional, but they had found themselves almost begging for guidance from more accomplished teams.

The effort and teamwork paid off, and their alliance stormed to a 92–34 victory. The D'Penguineers finished the day ranked tenth out of the eighty-six teams in the Galileo division.

"That match was lovely," Chase said. His bout of nerves had passed, but he feared another one tomorrow. "What's nice is when we pass people on the way to the stadium and they go, 'That's 1717.'"

As they rolled the PenguinBot out of the stadium, they heard the emcee announce over the loudspeakers, "Get some rest, teams. We'll crank it up again tomorrow, 8:40 A.M."

At restaurants and hotels across downtown Atlanta, teams were far from relaxing after the intense day of competition. Now it was time for scouting sessions in anticipation of each division's alliance selections.

In an all-night diner with orange beams running across the ceiling and lime-green tables, Amir sat with his drivers and scouts. "You've got to talk to HOT," Stuart said, looking over at Amir.

Andrew said, "We want them as a teammate, without a doubt."

"HOT wants to win," Amir said, adjusting the baseball cap on his head. "In Michigan, they went with the ThunderChickens because they knew they could win with them, because the Thunder-Chickens know how to coach and they have experience. If they don't pick us, we need to go with a team that can shoot and has the ability to own other drivers."

With every minute of the conversation, everyone at the table looked more and more exhausted.

"I wouldn't pick WildStang," their pit boss, Daniel, said. "What can they do that we can't do?"

"They score about a hundred balls at a time," Andrew said.

WildStang, a Motorola-sponsored team out of Schaumburg, Illinois, had a massive bin able to store almost thirty balls. They dumped them in a fast continuous stream between two rollers. Their robot also featured a swerve drive that made their robot as maneuverable as the PenguinBot. Undefeated that day, they were ranked second behind HOT, and everybody at the diner that night feared the two would pair up to create a super alliance.

Both teams had power dumpers. Having now participated in or watched hundreds of matches this season, the D'Penguineers knew this was the most effective design for Lunacy. Accurate, efficient scoring was done right up against a trailer, and these robots unloaded a huge volume of balls at once. The versatility of the PenguinBot with its omnidirectional drive and vision-targeted turret made it almost as effective, but not quite.

Amir said, "You guys need to prove to me that Turk plus the PenguinBot is better than WildStang. Then I can go to HOT."

After dinner, they returned to the motel, where the scouts met in one of the rooms until well past midnight. Down the hall, Amir tried to stay away, still feeling ill from the flu, but he kept knocking on their door with a new thought.

"I'm going to bed for good," he said after his third visit. He shut the door but returned ten minutes later. "This is the nugget of truth. We shut down teams because of the nature and style of our play. We make teams do what they don't want to do. That's what we need to tell HOT."

"What's the nugget?" Yidi asked.

"That's more like a chicken finger," Stuart said.

Everybody was getting slaphappy.

Amir slammed the door in mock frustration. Three minutes later he was back yet again. He wanted to see the data of the different teams broken down on a spreadsheet, but the scouts hadn't entered anything into their computers.

"It's the human mind," Stuart said, tapping the side of his head. "It's like the 1950s."

Finally, Amir retreated to his room for the rest of the night. His students, all of them coming together as one, had done a masterful job of managing the pit, presenting to judges, working on the PenguinBot, and scouting their division—almost without his guidance. This was their team now. They owned it. Their choice of alliance partners, whoever they were, would be perfect because it was a decision they had earned the right to make on their own.

"The big day. This is what decides it all," said a grizzled coach in army-green pants to his team while riding down the elevator in the Super 8. It was Saturday, April 18. Championship day.

The doors opened onto the mezzanine level. D'Penguineers in their black flight suits already filled the tables. For 6:30 A.M., they were a lively bunch. After the months of hard work, they had grown used to the late nights followed by early mornings. They shoveled cereal into their mouths and gulped down juice. A few held Styrofoam cups of coffee or tea. They had only a few minutes before they had to file out of the motel and march toward the stadium. All of them knew that their many months of effort would come down to how they performed this final day.

On the walk over to the pits, Gabe and several others avoided talk of the matches ahead. Instead they debated whether or not *FIRST* had a cool factor.

"There's some serious *uncool* walking around this place." Gabe laughed, reminding them of some of the crazier costumes he had seen during the past two days.

"It's cool because our whole school thinks it's awesome. We've had some cheerleaders come to our competition," Kevin said.

After giving it some thought, Chase added: "I don't know if it's that robotics is cool, but whenever you do something so intensely and go all out, it's pretty much cool, no matter what it is."

Their mentor, Danny, walked alongside them and spoke about how a robotics team was no different from high school sports teams that traveled to competitions. By contrast, one didn't see theater groups travel, he said.

Gabe said that he knew high school theater groups that went on the road to put on shows.

"Uhhh, theater isn't cool," Chase joked.

"What did you say?" Gabe swaggered like a tough guy. "What did you say?"

"What people don't get," Kevin said, "is that the product of all this work is a massive robot. Normally hard work ends with 'Oh, Gabe got an A on his math test. Good job.' But they see this and say, 'Wait, high school people do this? Are you kidding me?'"

"You'll remember this as cool," Danny said. "When I played football, the year we won state, I remember *every moment* from that year. This season will stand out like that for you."

They reached the World Congress Center and waited for the steel shutters leading into the pits to rise. Students amassed there like restless cattle in front of a gate.

When the shutters rose, everyone rushed through to get to their robots, run systems checks, and strategize for their final two qualification matches before alliance selections.

With his fellow programmers Kevin and Nick, Gabe rolled the PenguinBot to the practice field to work on their autonomous yet again. Over the loudspeakers, the pit area announcer said, "Can I have your attention, please?"

"The voice of God," Kevin said, looking up.

"Team 870 would like to have the Alan wrench that they shared yesterday with another team to please be returned now. It was a lender, *people*. Not a gift."

"Wow, God's kind of a jerk," Nick said.

Then they got to work.

Back in the pits, Turk attempted to psyche Chase up for the day. "We're ready to go undefeated today, right?"

Chase exaggerated a yawn. "Turk, we'll talk in like ten hours."

"Why does it have to be like this, Chase?" Turk said.

"I want to be Zen," Chase said.

He was serious. He needed to stay calm and level. The night before every other competition day, Chase had suffered nightmares of driving the PenguinBot disastrously. The previous night, he'd slept soundly, no nightmares. He felt like he was as ready as he would ever be.

Amir arrived in the pits at 8:30 A.M., surviving off a regime of flu medication. After a quick scouting meeting, he met with their first alliance and then checked on the robot.

Gabe had the PenguinBot on blocks in their pit again so he could run each autonomous routine and make sure the wheels were spinning and turning in the direction they had recorded. The routine they needed for their first match in less than a half hour was not working.

"Do you trust any of these?" Amir pointed to the sheet where Gabe had drawn all of the movements of each autonomous routine.

Gabe was panicking, trying to figure out what was wrong while fielding questions from Amir. Some of the routines had not saved properly. He didn't know why.

They decided to use a different routine than they wanted and hurried to tell their alliance of the change in positions. Gabe sat down on a toolbox, his lips trembling, unable to speak. It was a lot of pressure to keep trying to improve their robot two matches away from the division tournament. Most teams were just holding their breath hoping that their robot would keep working.

In the sixth qualifying match, the PenguinBot moved in autonomous to their alliance partner's corner. They loaded up their helix, and when Chase took control, he made a quick assault, running up a lead. With a minute left, their opponent robot, Shred, built by Liberty Robotics (Team 2775), pinned the PenguinBot against the wall. No matter what traction-control mode Chase switched to or joystick movement he made, the PenguinBot failed to break free. For the rest of the match, they were shut down. They squeaked by with a 56–54 win.

This was the first time in competition that they had been unable to escape a pin, yet Amir and his students were worried that other teams in their division might now think the Penguin-Bot could be shut down. Still, it was a win, and now they were in seventh place in Galileo.

As soon as Amir returned to the pits, Andrew and Stuart gave him the data he had requested the night before about their team's scoring ability, both with the PenguinBot and Turk. His scouting captains wanted him to pitch their team to HOT. Reluctant to go on a sales call, Amir hesitated but then strode away. If HOT chose them in alliance selections, they were on as sure a path to victory as existed.

"Our robot scores thirty points on average, our human player sixteen," Amir said to their coach, Adam. "You're the only team scoring more than us."

"We're going with WildStang as our number-one choice."

"Good to know," Amir said. He liked straightforward.

"But you guys were right behind them in our mind."

Amir relayed the news to his team. To reach Einstein, they would have to defeat an alliance with two former champion teams with lots of experience, veteran drivers, and two of the highest-scoring robots in all of *FIRST* this season.

In their final qualifying match before the division tournament began, the D'Penguineers were paired with another high-scoring robot and won 90–64. It was another great show in teamwork. Amir and the drive team lifted the PenguinBot off the field and hurried back to the pits, feeling good. They'd finished 6–1 and ranked seventh overall in Galileo. They would largely be able to choose their own fate at alliance selections.

The half hour before Stuart had to walk onto the Galileo field was a whirlwind. In the stands, the team's scouts compiled their final data. On their sheet of preferred picks, Stuart and Andrew scribbled, crossed off, reordered, and added various team numbers. The two also traded cell phone calls with a number of teammates, trying to locate Amir. He never seemed to be in one place more than a few minutes.

At last, Stuart and Amir met at the entrance ramp to the stadium. Nearby, other coaches and their alliance captains were deep in conversation. During the draft process, *FIRST* allowed only students on the stadium floor. "Hey, Shaeer," Stuart said.

"What?" Amir asked while on yet another call, this time instructing his pit crew to gather their toolboxes, batteries, chargers, cords, and spare parts and bring them to the stadium. Once the tournament started, they could not return to the pits.

When Amir hung up, Stuart smiled at him and said, "It's all me."

"Yeah, it's all you. So what's the plan?" Amir asked, thinking about how in an unexpected turn of events HOT had lost their last qualification match, leaving WildStang with the number one rank.

"We're looking to get two robots that can score. Hopefully they can play some defense too."

Stuart and Amir looked at data on the other teams in the top eight.

"WildStang will pick HOT, we know that," Amir said. "That leaves us the odd man out. There's no way we'll find another robot anywhere near either of those two. Team 1318 is ranked second.

They'll probably pick us. The question is, do you believe we can do better than them?"

They knew that if they rejected an invitation from a higher-ranked team to join an alliance, then, by rule, they couldn't choose or be chosen by any other top-eight team. They would have to take their chances by picking two teams from the pool of seventy-eight left in their division.

Finally, Stuart strode out onto the field. The emcee gave him a long look as if to say, Who is this kid in the black flight suit with a wild mane of hair like an eighties rock star? On the three other fields in the stadium, alliance selections were also under way.

As expected, WildStang picked HOT.

Next the number-two-ranked Issaquah Robotics (Team 1318) invited the D'Penguineers as Stuart suspected they would, but he declined. The Washington State team was undefeated, but they had benefited from an easy qualification schedule. Stuart thought he could do better by making two selections himself, especially since the D'Penguineers would pick their third alliance partner before the higher-ranked teams because of the serpentine drafting process.

By the time he held the microphone to make his first selection, the top four teams on his list had already been selected. Number five on his list was Team RICE (870). Ranked thirty-fourth at the end of qualifications, the New York team had suffered from a tough match schedule, but their robot's dumper scored an average fifteen balls per match. The data did not lie.

Stuart asked them to join their alliance. They accepted.

A few minutes later, Stuart went with an obscure choice for his second alliance pick: Liberty Robotics (Team 2775). A first-year Tennessee team, they ranked thirty-first at the end of qualifications. Their robot, Shred, with its simple frame and six-wheel drive train, had pinned the PenguinBot in their sixth qualifying match. Although the team was 4–3 in qualifications, their robot averaged eight balls a match with its elevator-like hopper. Their driver had also proven able to shut down his opponents.

When Stuart uttered their number, the Tennessee team leaped to their feet in the stands and started jumping up and down in their yellow-and-black outfits. They were the only rookie team selected for the Galileo tournament.

With these two midranked teams, the D'Penguineers would have to defeat the best of the best in their division to reach Einstein. The juggernaut of the HOT and WildStang alliance loomed large in their minds. Amir and his students felt like their season would be defined one way or the other by their ability to defeat this powerhouse. It would not be easy.

"Let's just keep this straightforward," Greg Needel, the mentor for Liberty Robotics, said to Amir before their quarterfinals against the alliance led by Hardwired Fusion (Team 708). "All we're going to do is pin one robot, dump on them, hold it, and let one of our other robots come dump. It's tag team all day. The only time any of us should release a pin is if someone's coming to score on your trailer, then you get out of there real quick."

"Okay," Amir said, delighted to discover the rookie team out of Tennessee had as their coach an eight-year *FIRST* veteran who knew what he was talking about. "We need some man-to-man too. Or I'm worried everyone will get in each other's way."

"All right," Greg said, specifying who would focus on which opponent. The twenty-five-year-old Black & Decker engineer with a baseball cap, long sideburns, and a goatee had once briefly worked at DEKA. His every movement seemed like a punctuation.

Amir instructed the human player for Team RICE on how

to load the PenguinBot in autonomous. Then he headed over to Chase and Kevin, who didn't even look up from the PenguinBot on his approach.

Turk rolled his shoulders. He was feeding off the energy of the whole stadium. Already at a fever pitch near the end of qualifications, the Georgia Dome had somehow reached an even higher level of crazed enthusiasm as the four division championships unfolded at once. Human waves rolled through the stands. Team chants melded into one tremendous roar. Students swung banners, danced, pounded their feet in unison, and shouted at the top of their lungs until their voices broke.

After the quarterfinal match before theirs finished, Chase and Turk lifted the PenguinBot onto the field. Gabe stood on the sidelines, his forehead beaded with sweat. As in the regional tournaments, each round was a best-of-three match format. Two losses, and they were packing up and finished.

The emcee introduced the teams. "And here they are, representing the West Coast. They're always on the hunt. Team 1717 from Goleta, California. They *are* the D'Penguineers!"

Amir and his drivers waddled back and forth. Before Team RICE was announced, their human shooter dropped to the ground, did a set of push-ups, jumped up, clapped, and then hopped up and down.

The buzzer sounded. In autonomous, the PenguinBot headed straight to the corner to load up. Gabe looked up at the stadium ceiling and exhaled. "It worked. Yes. Yes. Yes."

Soon after Chase took control, the PenguinBot unleashed on their opponents, racking up a huge lead. Chase set up another explosive shot, Turk went 10–13, and their alliance-partner robots and human players added their own scores for an overwhelming victory: 102–50.

After they brought the robot off the field, Gabe and Kevin performed a quick systems check. "We're good," they said. Amir and Stuart studied the scouting report on the match.

"I'm so happy with our alliance," Stuart said. "They did really well, and nobody knew who our teams were."

Before their second match, there was a long wait. Thoughts of the academy never far from his mind, Amir told Greg about his four-year program and his ambition to establish similar academies across California and the United States. "I'm trying to change the world," he said without a hint of bashfulness.

Chase, Kevin, and Turk stood side by side, quiet as they watched the other quarterfinals. The HOT alliance rolled through theirs, while Team Hammond, surprisingly, lost.

"I'm in my Zen place," Chase said, his hand on the Penguin-Bot before their second quarterfinal. He was oblivious to the CBS camera crew encircling the robot.

Before they set up on the field, Amir called Gabe over to ask about the autonomous routine they were using for this match. They had never tested it. "What's your level of confidence it'll work?"

"We're good," Gabe said, though he was sure he wouldn't be able to watch the match during the first fifteen seconds.

They were good. Their alliance won 88–66 to advance to the semifinals.

Now they would face the number-two-ranked Issaquah Robotics, who were paired with the eleventh- and twelfth-ranked teams in their division, the Greybots and Raider Robotix, respectively. The D'Penguineers knew the Greybots too well. Their robot, Raptor, had scored on the PenguinBot in the final seconds of the Los Angeles Regional to claim the title.

"We need you to shut down the Raptor," Amir said to his alliance. "They're fast but don't have a lot of traction."

"Shred will be on them all day," Greg said. "Know that the Raiders' Evil Machine will come for you."

"Pinning Raptor is like pinning HOT. Think of this as practice."

"We're leaving everything out there," Greg said.

"I want to fill in autonomous again," Amir said before turning to Stuart. "But if we do that, will we get killed by another human player as we load?"

"All three of their shooters are average," he said.

"Then we'll do it," Amir said. He walked away from the huddle to find Chase. "We move and score. Duck and dive."

A time-out by their opponents delayed their first semifinal match. Amir met with his alliance partners again. "My scouts tell me Issaquah is easy to pin. We'll go after them. The Evil Machine will be following us. Team RICE, you follow them to score because they'll be exposed going after us. It'll be a big train of stupidity." He smiled.

The emcee called the end of the time-out. He was skating about the field on Rollerblades, wearing purple pants and a flowered shirt. When he stepped off the field, he boomed, "We're green! Drivers behind the line! Ready. Three, two, one, go!"

In autonomous the PenguinBot moved to the corner where their alliance partner's human player began rolling balls into its helix.

Autonomous ended, but Amir urged the student to keep feeding balls until the moment before their opponent tried to pin them. Chase directed their robot out of the corner. Thanks to Turk, who was 7–7 from his seated position, their alliance was up 30–14 in the early part of the match.

"Perfect. Perfect," Gabe said from the sidelines.

Shred hounded the Raptor, which had a full load of balls in its hopper. But their opponent's driver was quick, and the Raptor circled their rookie alliance partner to score. At the last second,

Shred managed to escape. The tempo of the match was much faster than any other they had played. There were almost no clusters of robots, and everybody was moving strategically about the field.

"The Raptor. Over there," Kevin said, nudging Chase with his elbow.

Chase swung his joystick to the left, and the PenguinBot responded, pushing the Raptor against the wall.

"Yes!" Gabe yelled. "They're locking down on them. Shoot!"

Kevin fired on the Raptor's trailer as Chase continued to pin them. But then the Evil Machine started unloading on the trailer hitched to the PenguinBot. "Get out of there!" Amir yelled.

Amir pointed toward another opponent robot, and Chase rotated the PenguinBot. Their automatic tracking system locked onto the trailer, and a stream of balls scored. With twenty seconds left, the Raptor had a bunch of balls in its hopper. "Go pin 'em," Amir said.

The PenguinBot tracked down the Raptor again and accomplished what their alliance partners had not: shutting their opponent down.

"Yes!" Chase and Kevin said together when the final buzzer sounded. They won 90–68. One victory away from the division finals. In the stands, their teammates shook their black-and-white pom-poms and screamed, "Blue Alliance!"

Before their next semifinal match, they watched HOT and WildStang roll to victory again, claiming a spot in the division finals as expected.

Amir and his alliance coaches decided to repeat the same strategy but to focus even more on pinning the Raptor.

At the start of the match, the PenguinBot again loaded up in autonomous, Amir calling to his alliance human player, "Feed the monster!"

Chase had trouble locking onto a trailer until the midway

point of the match, when he zeroed in on the Raptor. Kevin fired continuously into the trailer. The stands went wild as the emcee said, "Seventeen-seventeen's scoring big-time into that trailer!"

Their alliance now had a thirty-point advantage.

The Raptor erased it with one strike. The robot escaped a pin set by Shred in the corner and then circled around for a full dump of balls into Shred's trailer. With thirty seconds left, the score was tied. The clock ticked down, the two alliances trading shots. Just before the buzzer sounded, the human player from Team RICE lofted a Hail Mary and scored into an opposing trailer far across the field.

The shot tied the score 84–84. They were on to a third match.

Before the match, Chase stared at the field, arms crossed. Kevin stood beside him, hands at his hips. They both looked like they were trying to hold themselves together.

In autonomous, human players for both the Raptor and the PenguinBot filled their robots in preparation for big scores, but the match turned into an all-defense game. Chase followed the Raptor wherever it went for the first half, scoring an occasional point but nothing big. That left them open to Issaquah Robotics, who scored on the PenguinBot trailer. Their alliance robots contributed a few points, but Turk and the other human players accounted for most of the scoring.

With thirty seconds left, the real-time score had the D'Penguineers leading by four points, 46–42. A dump by the Raptor, still full of balls, would end the match. Then Shred caught the Raptor in the corner and pinned it. Their opponent was going nowhere.

The D'Penguineers and their partners scored several more balls. With twenty seconds on the clock, their opponent tried to loft a super cell but missed. The emcee said, "Super cell on the field." A second later, the Evil Machine picked it up.

"They have a super cell," Amir warned.

But it was too late; the Evil Machine had already followed the PenguinBot toward the corner. They were lined up for the shot, the super cell the only ball in their robot.

"Are they going to score?" the emcee asked. "Five, four, three..."

The super cell spun in place inside the Evil Machine. It was stuck. The buzzer sounded.

"Ladies and gentlemen," the emcee said. "We have our second finalist. The Blue Alliance wins 64–46. They'll go to the finals!"

Amir raised his arms overhead. "Yes!"

Chase and Kevin hugged, and their alliance partners all high-fived one another on the field.

Immediately, Amir started strategizing how to bring down the two dangerous scorers HOT and WildStang in the Galileo final. Their alliance partner, Spartan Robotics out of California, was not to be dismissed either. They had finished the division qualifications ranked sixty-sixth, but this was a deceptively low ranking for a team who had gone undefeated at the Silicon Valley Regional and whose robot scored an average ten balls a match.

"You guys need to go after HOT," Amir said to the Shred's drivers. "We'll try to shut down WildStang. If we can lock both of them up, that leaves Team RICE to duel it out against Spartan."

"We'll do our best," Greg said, knowing the Everest they had to climb.

A few minutes later, Chase and Turk prepared the Penguin-Bot on the field. On the projection screen, the cameras zeroed in on the gold-painted shooter, then on the blinking lights of their well-organized electrical board.

The emcee introduced the alliances. "Ladies and gentlemen, welcome to the finals of Galileo. We have some years of experience over here in the Red Alliance. They're in the Hall of Fame for good reason, Team 67, HOT. Also in the Hall of Fame, you know the tie-dye, you love being with them, Team 111, Wild-Stang. And a California West Coast power, Team 971, Spartan Robotics!"

He then turned toward the other side of the field. "Down in blue, it's a Cinderella story. From New York, that beautiful machine has been in the corner all weekend, 870, Team RICE. Our rookies Team 2775, Liberty Robotics. Wearing the flight suits, it's black and white, and it *is* pretty: Team 1717, the D'Penguineers. One of these alliances will represent us on Einstein."

"We're here," Amir said, holding his drivers close.

The buzzer signaled the start of the Galileo Division finals. Anticipating the PenguinBot to move to the corner to load up on balls in autonomous, HOT blocked them midway. This allowed the Detroit team's human player to score on the PenguinBot trailer and threw Amir and his drivers off their game plan. When autonomous ended, they were down 12–8.

Throughout the match, HOT stuck to the PenguinBot, shutting down their strategy to go after WildStang. Team RICE tried to evade the number-one-ranked team but soon lost that battle. With a swerve drive they had perfected over many years of competition, WildStang maneuvered into place and unloaded.

With a minute and a half left, the D'Penguineers were down 40–30 on the real-time scoreboard, but the officials looked behind in their tally given the number of balls dumped at once by Wild-Stang. HOT and the Spartan robot also continued to rack up points. The PenguinBot and their alliance robots scored here and there but never more than a few at a time. Amir was silent with his drivers, nobody sure how to overcome these three effective offensive robots. At the very end, HOT scored seven balls into the PenguinBot trailer.

An easy win for the number-one-ranked alliance: 106–76.

"I don't want to go home now," Turk said, looking up at the stands before he repositioned the PenguinBot.

Amir searched for a winning strategy with his alliance partners, but he knew the odds were against them.

HOT called a surprising time-out. Amir learned that the intake roller on their robot broke in the previous match when they had stopped the PenguinBot in autonomous. Six minutes later, the HOT team had fixed the problem. In that time, Amir had reconsidered their strategy. The D'Penguineers would go after WildStang as they had planned before and shut them down. Given how many balls they had dumped on Team RICE, Amir thought there was no way they could win unless the PenguinBot took out WildStang.

The competing alliances circled the field to shake hands, further reminder to Amir and his drivers that they were a loss away from elimination.

The second finals match began. This time WildStang cut off the PenguinBot in autonomous, but Turk and their other alliance human players scored an impressive twenty points in the period to take a 20–6 lead. When Chase grabbed the controls, he focused on WildStang, and Kevin unloaded seven balls into their trailer. HOT made up several points with two targeted bursts of shots. Then WildStang followed up with a dump. Team RICE stretched the lead again for the D'Penguineers alliance when all six robots clustered in the corner.

"Lock WildStang down!" Amir implored Chase.

Chase directed the PenguinBot over to the side of the field and pinned WildStang against the wall. Their omnidirectional drive train was proving more than equal to the task of going against the famous WildStang swerve drive.

"You own them, Chase. You own them," Amir said.

But with the PenguinBot no longer playing offense, the team was relying on its partners to score. Shred and the Team RICE robot managed a few more points, but were nowhere equal to the firepower of HOT, which now had free rein on the field. They scored burst after burst. The score evened, then the HOT alliance took the lead and widened it.

In the few remaining seconds, Team RICE ran an empty cell to the corner, where their human player waited to exchange it for a super cell. This was their last chance at coming back. Shred baited an opposing robot over toward Team RICE's player in the final seconds. He threw the super cell, but it bounced off the trailer pole and fell to the field.

The horn sounded.

A few moments later, the emcee said, "Ladies and gentlemen..."

"It's over," Amir whispered.

"...moving on to Einstein, it's the Red Alliance with a 104–88 victory!"

As HOT and their alliance partners leaped up in celebration, Team RICE's human player dropped to the floor and held his face in his hands, his whole body trembling. Amir went over and placed his hand on the student's shoulder. "The shot wouldn't have mattered anyway. You just got second place to the alliance that'll probably win the whole thing."

The kid shook his head, unable to find words.

Gabe kneeled down on the sidelines, his face rigid, fighting back tears. Turk, Chase, and Kevin took the robot off the field and came over to his side. They would not be going to Einstein.

"We can't ask for anything more," Amir said, standing before them. "We've nothing to be down about. We were the seventh seed, we picked our own alliance, and we almost did it, even against an alliance that made it impossible. We proved we deserved to be here. No one had seen us play defense before. Our drive train showed how awesome it is against WildStang."

The HOT coach, Adam, came over. "You did a really great job, all of you."

"We'll be rooting for you," Amir said. "You deserve to win. Your alliance should take the whole thing."

"Thank you," he said before walking away.

"We finished top eight in the world," Amir said to his students.

Gabe lay down on the stadium floor beside the PenguinBot, his eyes on the luminescent dome above. Chase did the same. Kevin and Turk were sitting in nearby seats, staring at the empty Galileo field. For a moment, the four of them remained still.

After four months of living and breathing robotics, after all the hours of work and intense devotion building the PenguinBot, they had nothing else to do and were sure they had nothing else to achieve.

"I look at the world like one of those pigs," Dean Kamen said, recounting the fable of the Three Little Pigs and their houses of stick, straw, and brick. Listening were twenty-five thousand students, mentors, and fans who had congregated on one side of the Georgia Dome to watch the four division alliance winners compete for the *FIRST* Championship.

On the edge of the Einstein field, Amir sat cross-legged on the stadium floor. His drive team and pit crew were crouched around him. They were all tired and hungry. They couldn't join their team in the stands and catch a bite to eat because the D'Penguineers might have to substitute for one of the teams in the winning Galileo alliance if a robot became inoperable. It was not how they wanted to make it to Einstein, but they needed to be there.

As Dean continued, Amir shifted on the floor and found a package of two granola bars in his flight-suit pocket. He slowly opened it, the crackle of the wrapper loud in the silence surrounding the speech. He broke off a piece of a granola bar for himself and passed the package to his mentor, Danny. Danny took a piece and handed the package on to Chase. Chase took the rest of one bar and passed the other one to Kevin, who broke off a piece, then on to Turk and finally Gabe. Not a word was spoken, or needed. They were a family now.

"Look at the world everybody has created. The houses of sticks and straw have come down," Dean said, playing out his metaphor about the pigs and an economy built on a housing bubble and Wall Street shenanigans instead of science and technology. "Everybody's worried about the big bad wolf except the little pig that built himself a really sturdy house. For almost twenty years, we've been building a strong house with *FIRST*. There's never been a time when what we're doing is more important. We need to go faster. We ought to be leading a lot of people, political leaders and others, in a direction that makes sense. Let's get back to an innovative, entrepreneurial, wealth-creating society that has world-class people that can solve ever more complex problems."

The stadium cheered when Dean finished, as much for his speech as the onset of the championship tournament.

In the semifinals between Galileo and Newton, the HOT alliance continued to dominate, winning both matches by almost fifty points each. The ThunderChickens, whose alliance was composed of all Michigan teams, had won Curie in a breathtaking three-match division final. Now in the Einstein semifinals against the Archimedes division winner, they won two straight. This set the stage for an epic rematch between the Ford- and

GM-sponsored teams. The year before, the ThunderChickens had beaten HOT to claim the championship.

Before the showdown commenced, Dean returned to the podium to announce the winners of the GM Industrial Design Award and the Motorola Quality Award. The judges, who were NASA engineers, research scientists with Ph.D.s, inventors, and professionals from Fortune 500 technology companies, had been circling the pits throughout the competition, taking notes and interviewing students about their robots to see who had the best-engineered designs. Given that *FIRST* was at its heart an engineering competition, teams highly coveted these two awards.

The Industrial Design Award, given to the team whose robot best met the challenges of the game, went to HOT.

Amir clapped, not surprised. The Detroit team deserved the award for their deadly efficient power dumper alone.

The judges awarded the Motorola Quality Award to the team who built the best robot in all of *FIRST* in terms of conceptual design, advanced fabrication, and machine durability. Its presenter started to describe the winner: "This team and machine displayed the highest degree of excellence and set a high standard for themselves and stood out as an example to other teams. We're proud to recognize this winning team as *the* quality standard for *FIRST*. This wayward team from the South Pole..."

Amir and his students looked at one another. They were all thinking in geographic terms, and Goleta, California, was *not* in the South Pole. Anyway, these engineering awards typically went to experienced, long-standing teams. Another second passed before it dawned on Amir and his D'Penguineers that there were no teams in *FIRST* from Antarctica.

The presenter continued, "...didn't swerve off course with their modular drive train. They Penguineered their way to success—"

Amir jumped to his feet, clapped, and yelled, "Yes!" Chase

and Gabe hugged and shouted. Turk threw his arms around all of them, and they bounced up and down together. In the stands, the rest of their team erupted out of their seats. The big projection screen showed the PenguinBot in competition, moving agilely about the field and firing balls with its turreted shooter. "Perfect Day" by Hoku boomed on the loudspeakers.

On the first day of their season Amir had told his students that winning the competition could never be guaranteed, but that they should aim to build the most advanced, professionally engineered robot. More than just validating all their work, the award proved that they had achieved their goal.

Stuart and Andrew hurried down from the stands to receive the trophy from Dean Kamen for the D'Penguineers. With his students surrounding him, Amir looked toward the illuminated dome.

"This means everything," he whispered, while the twenty-five thousand fans in the Georgia Dome applauded his team for what they had achieved in building their black-and-gold PenguinBot.

Epilogue

I t's a coin toss," ThunderChickens coach Paul Copioli said before the *FIRST* Championship finals began. "And I hate coin tosses."

Across the field, HOT coach Adam Freeman looked at Paul and raised his arms overhead, signaling that he had no tricks up his sleeve. Paul did the same, even though he was wearing his fluorescent green ThunderChicken T-shirt. The two teams had been allied throughout the season in Michigan. They knew each other's strengths and weaknesses. They knew each other's strategies in autonomous and the driver-operated period. They figured their own robots were equal to each other. The question now: Who had assembled the best alliance and best executed their strategy to win?

As the emcee chronicled the storied history and many interconnections of the six teams in the Einstein final, the drivers stared at the field or down at their feet. They seemed oblivious to the enormity of the stadium, to the long sweep of fans cheering in the stands, and to the cameramen and photographers capturing their every move. They were only thinking about what they had to do.

Then the buzzer rang. The match started, and it was on. In autonomous, the ThunderChicken robot, Roxxy, made a long arc to reach their corner to fill up with balls, avoiding a pin that Adam had planned to set. He was surprised by the move, since the ThunderChickens had never done that before. Even after playing more than ninety matches this season, Paul still had a few tricks left.

The battle was an offensive free-for-all. By midmatch, the score almost tied, few balls remained on the field; they were either already in trailers or in robots angling around for a score. HOT and their alliance partner Spartan Robotics had the lion's share, and both managed to unload them. Only by scoring super cells would the ThunderChicken alliance have a chance now. The HOT driver Nick Orlando shut down Roxxy in the final twenty seconds and prevented a super-cell shot. Another dominant win for the HOT alliance: 100–68.

HOT had the ThunderChickens outgunned.

Two minutes and fifteen seconds later, Nick Orlando was leaping into the air and hugging his coach, his teammates, and his alliance partners. With a 98–81 win, they were the 2009 *FIRST* champions. Streamers fell from above, and "We Are the Champions" played throughout the stadium.

The two Michigan teams met in front of the Einstein field, congratulating and clapping one another on the back. Many of the students were friends now.

In the stands, Gabe Ruiz and Adam Cohen, members of 2Train Robotics, sat side by side. The New York team had found themselves struggling in almost every qualification match. They'd finished 3–4 and had not been drafted for the division tournament. None of that mattered now. Both were off to college in the fall. Both would be forever changed by their experience on 2Train.

Over the next hour, the Georgia Dome emptied of thousands of high school students and *FIRST* fans. Work crews brought down the curtains and disassembled the regolith fields. Soon enough there would be no sign that a robotics competition had ever taken place inside the stadium.

In the World Congress Center, teams packed up their gear, broke down their pits, and lifted their robots into their crates one last time for shipment back home.

"This is what winning looks like," Amir joked to his D'Penguineers, referring to all the work they still had to do at this late hour. They gathered their things and started the trek back to their motel, hauling the hundreds of pounds of tools and supplies they had brought to the competition. Before he left, he patted the side of the crate holding the PenguinBot. "See you later, pretty baby."

That night there were no long speeches back at the motel. Amir congratulated his team and told them how proud he was of everything they had accomplished. Then it was all schedules and to-do lists to prepare for flying everyone back to Goleta. Chase sat in the back, wearing the same satisfied, easy smile that he had worn since the competition ended.

Gabe sat beside him with the same smile. For the first time in more than three months, he had no programming to do. He thought proudly of the day before when a representative from National Instruments had approached him, both amazed and curious at how Gabe had accomplished everything he had with his automatic tracking system on their turreted shooter. These were the people who had made the robot controller, and they were asking for *his* guidance.

Once back in Goleta, the D'Penguineers were fêted by their Dos Pueblos classmates. At a school assembly, the team showed off the PenguinBot, everyone in awe of what it was able to do. They wore their flight suits and the medals they had won on campus. They felt cool, decidedly cool.

Being on the robotics team even helped score a date for John Kim, Amir's student aide. One afternoon, he made a sign that read, KAREN, WILL YOU GO TO PROM WITH JOHN? He then decked out the PenguinBot with flowers and had Chase drive it up to the classroom where Karen was sitting. Chase shot balls through the doorway, getting everyone's attention, and when she came to look, Chase spun the robot around, presenting John's sign. She said yes.

After the long season, everyone was glad not to spend their every waking hour outside class in the build room. Gabe finally had time to spend with his father, who was recovering from his bone marrow transplant. But they soon found they missed the intensity of the experience too. Chase, who was now set to enroll in the mechanical engineering program at UCSB, hoped his future classes would offer at least a sliver of the same.

When friends complained of how much studying they had to do or how little sleep they were getting because of this club or that class, the D'Penguineers could not help but smile. They had redefined for themselves what hard work was all about. They knew now, as never before, the commitment one needed to achieve excellence.

Almost to a student, the members of his 2009 academy senior class believed that they would look back at their time as a D'Penguineer as the most transformative of their high school experiences, whether it led to a career in a STEM field, taught them what it took to be the best, or gave them a sense of what it was like to feel cool for using their smarts.

Once the *FIRST* Championship concluded, Dean Kamen made his way to a private facility at the Atlanta airport where his plane was waiting for him, already fueled and outside the hangar. His

Hawker Beechcraft Premier could fly at an altitude of more than 40,000 feet and was the fastest nonmilitary single-pilot aircraft on the market.

He was accompanied by his chief of staff, Kerri Maxwell, who had, among other duties, the Herculean task of managing her peripatetic boss's schedule. They walked straight onto the plane. Dean ran through his preflight checklist, and a few minutes later they were taking off into the sky, the theme music from *Star Wars* playing in their headphones. Dean, who had been flying for more than thirty years, never took off without it. In two hours, they would be landing back in Manchester. There was no time to waste in the clouds.

For the past two months, he had zigzagged across the country, attending competition after competition. He gave speeches at opening ceremonies, walked through the pits to talk to students and mentors, and watched scores of matches, proud of what these kids were able to achieve.

Everywhere Dean traveled throughout the season and beyond, he spread the message of *FIRST*. He recruited corporate sponsors and prodded long-standing ones to do more. He pushed for new teams to be formed, more mentors to contribute their time. He petitioned every politician he could to help support the organization in their home states or towns. He mentioned *FIRST* in almost every media interview, even when it seemed way off topic. Incorrigible was not a personal description he minded. "When I'm in a room, people around me hear a sucking sound," he told a *Popular Science* reporter. "That's me trying to take any bit of energy they have left and redirect it to *FIRST*."

After eighteen years, even with the phenomenal growth of *FIRST*, Dean knew he had not yet changed the culture. Professional athletes and movie stars were still the biggest celebrities, and American kids were still not pursuing STEM careers at anywhere near the level needed. The country needed a new generation of inventors, scientists, engineers, mathematicians, and technologists as never before.

The American economy continued to suffer from a contraction that many were now calling the Great Recession. Industries throughout the country continued to cut back on production. Unemployment was spiking in every state. Foreclosures still ran rampant. Household net worth was taking a beating. Federal and state government deficits were ballooning. Wall Street was only beginning to stabilize after a massive infusion of public capital. Prospects for recovery looked bleak and distant.

Dean Kamen remained sure as ever that he knew the solution to seeing the country through the crisis: *FIRST*. Later that year, he brought his message straight to the White House, standing beside President Obama at the launch of the $260 million "Educate to Innovate" campaign to partner the government with companies and foundations in engaging students in science and math. In attendance at the event was a team with their robot from the Lunacy season.

Obama was not the first president Dean had stood beside who promised to devote himself to inspiring students to pursue STEM careers. No doubt Obama was sincere, but Dean didn't plan to depend on him or anybody else to realize his vision of having *FIRST* in every high school in America. He would see to it on his own, leading the charge.

But as always, he felt more needed to be done, and faster, both for America and the whole world. After the recent death of his father, Dean was aware that at fifty-eight he himself did not have forever to achieve this and everything else he wanted.

"Other people have peace because I think maybe they just don't think about it, the ticking clock," he said one late evening in the quiet of his wine cellar at Westwind. "But it's always there with me. I know that I have this finite amount of time, and there's so much to do, so much in this world that needs fixing. What if I don't have time to finish all this? I can't cheat death. I'm not a religious man, but I have a kind of faith. Faith that these kids will be able to finish what we've started."

At 4:30 A.M., Saturday, January 9, 2010, a white Toyota Matrix was alone on Highway 101, heading south along the Pacific coastline. Amir was at the wheel, traveling to Los Angeles for the kickoff of another robotics season. Like the year before, he was accompanied by two students, but this time Emily was back at home, tending to their four-month-old daughter, Aliya. Amir broke the silence to offer his lesson on how he could predict upcoming bumps in the road. It was a new season, a new class of seniors, a new team to create from a bunch of robotics rookies.

Since the success in Atlanta the previous April, Amir had made great strides in his ambition to expand the engineering academy. In the previous nine months, featuring the PenguinBot in all its glory, he and his foundation had raised $1.35 million, increasing their total funds to a little more than $2 million. One million came from a single donation by Virgil Elings, a former professor at UCSB who founded Digital Instruments. Other donations came from companies such as Raytheon as well as the fund established for Lindsay Rose. Amir had promised her parents that he would name the mechanical engineering lab in their new building after her.

By the start of the 2010 season, Amir was on the homestretch to raising the three million dollars he needed in matching funds to receive his grant from the State of California. Because of state budget issues, the deadline to raise these funds had been pushed back a year from November 2009. He would surely make that deadline now.

Soon his academy would be reaching more students, with a richer range of experiences, across a broader spectrum of academic standing. He would also start bringing in teachers to guide them on launching similar academies across California and beyond, revolutionizing how science and engineering is taught in America's classrooms. He wanted *FIRST* to remain a key element of his academy. Further, he believed that with the competition integrated into a school's curriculum, so that students received academic credit for their participation on a team, and teachers earned a salary for leading them, Dean Kamen's ambition of a robotics team in every high school would be realized.

Back at Dos Pueblos after the kickoff in Los Angeles, Amir met with his new team and their mentors, including Stan Reifel. A number of his 2009 graduates were there as well. On his winter college break, Gabe was among them, eager to help Amir out for the first few weeks of the season. The former D'Penguineers programmer advised one of the seniors that morning, "The mechanism that you think is impossible to build—it's not. The second you believe that, someone will show up at the competition with that ability. So just go for it."

A few days later, Gabe sat down with Amir in the build room after everybody had left. Gabe told Amir that he was enjoying the theater program at New York University, but he felt like he might be on the wrong path: Theater was a passion but not the only place where he wanted to make his mark now. His time on the robotics team had shown him that a career in computer science did not mean he'd be stuck in a cubicle, punching in code, for the rest of his life. The season had also proved to Gabe that his skills and creativity as a programmer could lead to infinite possibilities. He now wanted to study computer science as well.

On May 20, 2010, Amir and his foundation learned they had finally raised the matching funds they needed for the state grant.

Two months later, Amir broke ground on the new academy building with a ceremonial gold-plated shovel.

One day Gabe, or another student Amir or *FIRST* touched, will invent something that will change our lives forever. The world will look back and wonder how it all started. They will find this story—or one much the same.

Notes

When I first set out to write about a season in *FIRST,* my plan was to profile three teams, each with their own unique aims and approaches as well as individual stories. This presented a challenge since I needed to be there at every moment, from the kickoff to the six-week build season to regional competitions, and, I prayed, the *FIRST* Championship in Atlanta, Georgia. I wanted to be in on every conversation and significant turning point. The only way to achieve this—short of cloning myself—was to embed a reporter on each team and then split my time between the three.

Thanks to my own crazy travel schedule, as well as the devoted efforts of Mike Traphagen (who shadowed Team 1717), Molly Birnbaum (Team 395), and Kelly Kozlowski (Team 217), I assembled a vast collection of detailed notes, interviews, and audio and video recordings that allowed me to bring each team to life. Every conversation, action, scene detail, and emotional context appearing in the book is based on this reporting and is not a manufactured recollection/conversation, as is a new trend in nonfiction. Some follow-up reporting on my own of HOT (Team 67) gave me the stories of four teams.

As with all best-laid plans, the intersecting, balanced profiles of three or four teams proved not to be the most effective way to present the story of *FIRST* and its impact on so many lives. Therefore, I focused on the D'Penguineers of Goleta, California, whose experiences in the 2009 season best reflected my intentions with this narrative. That said, a book could probably be written on each team I followed, and I thank each and every student and mentor who contributed their time to this project. Many of their individual stories may not be told here, but the context they offered was invaluable to the final work.

I am also indebted to other journalists who have written on Dean Kamen and *FIRST*. Below I have sourced their articles and other research as well.

THE KICKOFF

12 **The hexagonal house:** John Richardson, "Lord Dumpling's Magical Water Machine," *Esquire*, December 1, 2008; Max Alexander, "Wow, Isn't That Cool!" *Smithsonian*, September 1, 2003; Adam Higginbotham, "Dean Kamen: Part Man, Part Machine," *Sunday Telegraph* (London), October 26, 2008; Nathan Cobb, "Mystery Science 2001 Inventor," *Boston Globe*, February 1, 2001; Devin Friedman, "Dean Kamen Wants to Change Your Life," *Ottawa Citizen*, September 17, 2001; Brad Stone, *Gearheads: The Turbulent Rise of Robotic Sports* (Simon & Schuster, 2003).

13 **Some likened Dean to Willy:** Higginbotham, "Dean Kamen."

14 **"Put a toggle switch":** Glenn Rifkin, "Profile: Dean Kamen," *New York Times,* February 14, 1993.

14 **He was always questioning:** Richardson, "Lord Dumpling's Magical Water Machine."

14 **"$F = ma$, that's":** Ibid.

14 **From that day forward:** Steven Levy, "Great Minds, Great Ideas," *Newsweek,* May 27, 2002; Cobb, "Mystery Science 2001"; Higginbotham, "Dean Kamen"; Friedman, "Dean Kamen Wants."

16 **"A good many times":** C. P. Snow, *The Two Cultures and the Scientific Revolution* (Cambridge: Cambridge University Press, 1998).

17 **"Albert Einstein, but":** Juan Williams, "Radio Interview with Dean Kamen," National Public Radio transcript, January 21, 2001.

18 **"philosophical love-in":** Rena Marie Pacella, "Dean Kamen Won't Be Satisfied Until He Reinvents Us All," *Popular Science*, June 5, 2009;

Brad Stone, *Gearheads: The Turbulent Rise of Robotic Sports* (Simon & Schuster, 2003).

19 **"I think it's time":** Dr. Woodie Flowers, "Educational Reform," speech at Olin University, May 9, 2009.

19 **"banana republic":** Rifkin, "Dean Kamen."

19 **He knew the statistics:** Bart Gordon, "U.S. Competitiveness: The Education Imperative; Because the Foundation for Future Success Is a Well-educated Workforce, the Necessary First Step in Any Competitiveness Agenda Is to Improve Science and Mathematics Education," Issues in Science and Technology 23, no. 3 (Spring 2007).

20 **"America's only going":** Thomas Donlan, "Robots to the Rescue," *Barron's,* May 2, 2005.

THE COMPETITION: THE THUNDERCHICKENS, TEAM 217

93 **The uncertainty over:** "The Year That Rocked Detroit," *Detroit News,* December 26, 2008.

93 HELP **a red spray:** "U.S. Car Industry in Crisis," *Guardian* (Manchester, U.K.), November 14, 2008.

94 OUT OF A JOB YET?: Ibid.

94 **"There isn't any light":** "City Tries to Hang On Amid U.S. Auto Industry Collapse," *Wall Street Journal,* April 6, 2009.

EPILOGUE

321 **"When I'm in a room":** Rene Marie Pacella, "Dean Kamen Won't Be Satisfied Until He Reinvents All of Us," *Popular Science,* June 5, 2009.

322 **"Other people have peace":** Ibid.

Game Glossary

alliance A partnership of three teams, which competes against another partnership of three teams in a match, whether in qualification rounds or an elimination tournament. In qualifying rounds, these alliances are randomly selected by computer.

alliance selection After the qualifying rounds, the top eight ranked teams draft two other teams that will be on their alliance for the elimination tournament.

autonomous period The fifteen-second period at the start of each match in which the robots follow preprogrammed routines instead of being controlled by the human pilots.

empty cell One of four orange-and-blue braided game pieces given to each alliance at the start of each match. The human player positioned at the center line of the field initially holds the empty cells.

They are worth two points if scored in opposing trailers, or the human players in the corners can exchange them for super cells.

field (crater)   A 27-by-54-foot field surrounded with a clear Lexan wall on which the matches are held. The diagram below provides a guide to the positions of the robots and players at the start of each match.

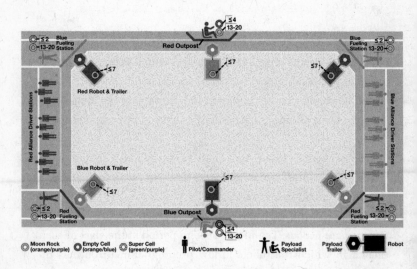

kit of parts   The basic frame, motors, controller, battery, wheels, camera, and scores of other components from which teams may build a fully operational robot. Provided to teams at the kickoff of the season, the kit is typically only the jumping-off point from which students and their mentors design their robots.

moon rock   One of twenty orange-and-purple game pieces given to each team to split between their human player and their robot. They are worth two points if scored into an opposing trailer.

payload specialist (human player)   The individual selected by each team who shoots game pieces by hand into opposing trailers. The three human shooters on each alliance are positioned in select spots around the field and must remain there throughout the match.

qualification rounds   The first part of a competition during which teams participate in a series of matches (typically seven to ten) with alternating alliance partners and opponents. At their conclusion, the eight teams with the best record are guaranteed a spot in the tournament as alliance captains.

regolith   The slick white material that covers most of the crater. Paired with specific robot wheels, the regolith creates a low-friction surface that makes it difficult to maneuver on the field.

robot pilots (drivers)   Two players on each team who operate the robot by wireless control, usually with joysticks. One pilot usually "drives" the robot, while the other operates its various mechanisms such as its shooter.

super cell   A green-and-purple game piece that may only be shot in the final twenty seconds of the match—and only if a human player in the corner has received an empty cell in exchange. Each alliance is allowed four super cells for exchange, and they are worth fifteen points if scored into an opposing trailer.

teleoperated period   In this two-minute period of the match after autonomous, the pilots command the movement and mechanisms of the robot by wireless control.

tournament   The format in which the top eight ranked teams form alliances and compete for the regional or championship title. Each round is a best-of-three match format. The winning alliance advances, and the loser is eliminated from the tournament.

trailer (payload trailer)   The 28-inch hexagonal basket attached to the back of each robot that serves as the target for scoring points (below). The pole topped with pink and green bands is the "vision target" that allows robots to track the trailer with its camera.

Acknowledgments

In all of my books, there is always a long list of people who contribute to its realization. Writing the first draft of a book such as *The New Cool* may be a solitary experience, but everything that came before and after is a collective effort. This book in particular required the generosity, patience, and time of a vast number of people. I thank each and every person involved (and apologize in advance to those I've absent-mindedly excluded).

First, my reporters, Mike Traphagen, Molly Birnbaum, and Kelly Kozlowski. This book would not have been possible without you. My apologies to all my late-night missives of what-I-need, and I hope you enjoyed this experience as much as I did with you. Thank you also, Gina Pace, who pinch-hit for me on one important occasion in Philadelphia.

At *FIRST,* my heartfelt appreciation goes out to the great Dean Kamen and his ever helpful chief of staff, Kerri Maxwell. Thanks also to Woodie Flowers, Cheryl Walsh, Jonathan Hawley, Jim Beck, Jon Dudas, Paul Lazarus, and Marian Murphy, the latter who initially shepherded me into this incredible community.

Thank you to my agents Scott Waxman, Farley Chase, Byrd Leavell, and Nick Harris. As always your guidance and skills made this book possible on so many levels. Also to Susan Mindell, who keeps me safe.

A big cheer to everyone at Crown, my publisher. Notably Rick Horgan, who first saw the potential with this project, and guided it with supreme skill into bookstores. Without his keen editorial insight, as well as that of his partners in crime, Julian Pavia and Nathan Roberson, I would still be lost a third of the way through the narrative. They reshaped this book at a critical time and are faithful evidence that great editing is still a necessary art in publishing. Kudos also to Julie Cepler, Jacob Bronstein, and Jill Browning. I also want to acknowledge Liz Stein, who helped me separate the wheat from the chaff after my first draft. Can't tell you how many trees she saved cutting out those many unnecessary passages.

Thank you to other early readers, including my nephew and former *FIRST*er, Tyler Roussos. Tyler first convinced me that this high school robotics competition was an untapped vein of narrative gold. Much appreciation also to his father, William, and the famous *FIRST* veteran Wayne Penn, who also provided critical reads.

To the *FIRST* teams I followed and profiled: There will never be enough words to acknowledge my indebtedness to all of you. Below I have included the names of each member. Thank you all. Some members suffered my attention more than others, and they deserve special note. Foremost, Amir Abo-Shaeer. His efforts to build his academy require endless devotion, yet he always made time for my endless questions and interviews during many, many months. His every success is deserved. Thank you also to his wife, Emily West, who sacrificed many weekends from her husband because of me. Also from Team 1717, Stanley Reifel, Sandy Seale, Gabe Rives-Corbett, Kevin Wojcik, Luke Seale, and Chase Buchanan provided much insight. From the ThunderChickens, I appreciate all the time from Paul Copioli, Neal Gandhi, Jason VanMaldeghem, Chelsea Fournier, and Ed Debler. From 2TRAIN, kudos to Bob Stark, Gary Israel, Marshall Fox, Adam Cohen, and Gabe Ruiz. Thanks also to Adam Freeman from HOT. As promised, a callout to each and every member of the four teams of *The New Cool*:

Team 1717: (Students) Tasha Bandeira, Daniel Berg, Chase Buchanan, Anthony Cazabat, Angie Dai, Isabelle D'Arcy, Max Garber, Saher Hamdani, Bryan Heller, Andrew Hsu, Daniel Huthsing, Caroline Kim, John Kim, Fedor Kostritsa, Jacob Kovacs-Goodman, David Liu, Lisa Nakashima, Alyssa Ogi, Matthew Parker, Colin Ristig, Gabe Rives-Corbett, Luke Seale, Stuart Sherwin, Anthony Turk, Nicholas Vaughan, Alejandro Veloz, Yidi Wang, Kristine Ware, Kevin Wojcik, Akifumi Yamamoto, John Yi. (Mentors) Amir Abo-Shaeer, Lennie Araki, David Boy, Ryan G. Conolley, Danny M. Lang, Juliana Bernal Ostos, Stanley Reifel.

Team 217: (Students) Ashley Arscheene, Jonathan Bezenah, Nicole Black, Ryan Brown, Eric Cai, Joe Chmielewski, Amanda Duffy, Joe

Ferro, Chelsea Fournier, Neal Gandhi, John Gurnow, Melissa Hartlage, Ian Hatzilias, Alex Henk, Jean-François Henry, Cameron Herringshaw, Ryan Jones, Anthony Kazyaka, Alex Lee, Claire Liburdi, Branden Mansour, Michael Maranzano, Christopher McMullen, Breanna Meyer, Sam Moore, Antonio Moraccini, Nischal Nakkina, Amanda Orban, Rebecca Pace, Joe Quesada, Daniel Rodak, Kyle Roeber, Jenna Ross, Hayley Schuller, Katharine Syms, Shaun Tobiczyk, Jason VanMaldeghem, Bradley Vogt, Kayla Wizinsky, Justin Wright, Joe Wyrzykowski, Evan Yahr. (Mentors) Ron Arscheene, Mike Attan, Bill Baedke, Mike Copioli, Paul Copioli, Ed Debler, Bob Korson, Anita Stafford, Gary Yahr.

Team 395: (Students) Nicholas Barker, Eric Castillo, Adam Cohen, Dequan Ellis, Daniel Espinal, Daniela Garcia, Mayra Garcia, Naya Gary, Sally He, Noah Kleinberg, Carmen Marrero, Brandon Oquendo, Bryan Pimetel, Jose Reyes, Alexandra Rojas, Gabriel Ruiz, Kevin Smith, Steve Thomason. (Mentors) Reuben Bridges, Kristian Bruno, Dan Cohen, Phillip Dupree, Heather Flay, Marshall Fox, Hans Hyttinen, Gary Israel, Henry Jones, Nima Karamooz, Tiffany Kwan, Paul Lucien, Bob Stark, Weade Williams, Crystal Zhou.

Team 67: (Students) David Armstrong, Kyle Barie, Brett Billington, John Bottenberg, Patrick Brennan, Christopher Brooks, Katie Broughton, Evan Cramer, Megan Crowley, Paul Doerr, David Drouillard, Derek Drouillard, Benjamin Elling, Gabrielle Elser, Alexandre Garrigo, Autumn Giles, Casey Grebe, Colin Hale, Ashley Haren, Kenny Harris, Jay Hendricks, Zack Herbst, Logan Laviolette, Ryan Legato, Elyse MacDougall, Candelaria Maxwell, Mitchell Maxwell, Kurtis McElroy, Joshua Orlando, Matthew Orlando, Nicholas Orlando, Andrew Parkanzky, Candace Ploughman, Gregorio Ponti, Laura Rarus, Stephanie Spurr, Alex Vogel, Jim Vogel, Garrett Wainstock, Rob Walkerdine, Denzell Walls, Jacob Wisnewski, Adam Zonca, Ethan Zonca. (Mentors) Less Arms, Bob Brines, Dave Doerr, Pat Doerr, Adam Freeman, Lori Gleason, Rodney Gleason, Samuel Grebe, Walk Hickok, Diana Houhanisin, Jim Meyer, Cindy Nader, Tom Nader, Theresa Prior, David Rubarth, Megan Rumble, Gina Sweet, Bob VanHam, Ken Velzy, Dave Verbrugge, MaryBeth Waldo, Rick Waldo, Jerry Wilson.

Finally, finally, thank you to my wife, Diane, for everything and more.

If *The New Cool* inspires you to learn more about *FIRST*®, please go to www.usfirst.org. There you can discover how to participate, whether it is to:

- **START** a team
- **MENTOR** a local team
- **SPONSOR** a local team, event, or overall program
- **FUND** a scholarship
- **VOLUNTEER** at local events

Although *The New Cool* features the *FIRST*® Robotics Competition for high school students, there are other fun and challenging programs for all ages of students, including

> Junior *FIRST*® LEGO League (ages 6 to 9; grades K to 3)
>
> *FIRST*® LEGO League (ages 9 to 16, 9 to 14 in the U.S. and Canada; grades 4 to 8)
>
> *FIRST*® Tech Challenge (ages 14 to 18; grades 9 to 12)

Further contact information: *FIRST*®, 200 Bedford Street, Manchester, NH 03101. Phone (603) 666-3906, (800) 871-8326; fax: (603) 666-3907

The *FIRST*® mission is to inspire young people to be science and technology leaders, by engaging them in exciting mentor-based programs that build science, engineering, and technology skills, that inspire innovation, and that foster well-rounded life capabilities including self-confidence, communication, and leadership.

FIRST®